C000213058

HOW MANY MOONS DOES THE EARTH HAVE?

Also by Brian Clegg

Dice World

Inflight Science

Introducing Infinity: A Graphic Guide

Light Years

Science for Life

The Quantum Age

The Universe Inside You

HOW MANY MOONS DOES THE EARTH HAVE?

THE **ULTIMATE** SCIENCE
QUIZ BOOK

BRIAN CLEGG

ICON

Published in the UK in 2015 by
Icon Books Ltd, Omnibus Business Centre,
39–41 North Road, London N7 9DP
email: info@iconbooks.com
www.iconbooks.com

Sold in the UK, Europe and Asia
by Faber & Faber Ltd, Bloomsbury House,
74–77 Great Russell Street,
London WC1B 3DA or their agents

Distributed in the UK, Europe and Asia
by TBS Ltd, TBS Distribution Centre, Colchester Road,
Frating Green, Colchester CO7 7DW

Distributed in the USA by
Publishers Group West,
1700 Fourth Street, Berkeley, CA 94710

Distributed in Canada by Publishers Group Canada,
76 Stafford Street, Unit 300
Toronto, Ontario M6J 2S1

Distributed in Australia and New Zealand
by Allen & Unwin Pty Ltd,
PO Box 8500, 83 Alexander Street,
Crows Nest, NSW 2065

Distributed in South Africa by
Jonathan Ball, Office B4, The District,
41 Sir Lowry Road, Woodstock 7925

ISBN: 978-184831-928-8
ISBN: 978-184831-999-8 (Book People)

Typeset in PMN Caecilia by Marie Doherty

Printed and bound in the UK
by Clays Ltd, St Ives plc

For Gillian, Rebecca and Chelsea

ABOUT THE AUTHOR

Science writer Brian Clegg studied physics at Cambridge University and specialises in making the strangest aspects of the universe – from infinity to time travel and quantum theory – accessible to the general reader. He is editor of www.popularscience.co.uk and a Fellow of the Royal Society of Arts. His previous books include *Science for Life*, *Light Years*, *Inflight Science*, *Build Your Own Time Machine*, *The Universe Inside You*, *Dice World*, *The Quantum Age* and *Introducing Infinity: A Graphic Guide*.

www.brianclegg.net

ACKNOWLEDGEMENTS

With many thanks to my editor Duncan Heath, and everyone at the excellent Icon Books for their support. Most of all, for every pub quiz I've ever taken part in and thought 'I wish there was more science'.

CONTENTS

INTRODUCTION

How Many Moons Does the Earth Have? has a traditional quiz format. The book contains two quizzes, each with six rounds of eight questions, plus two 'special rounds' which earn up to ten points and involve themed questions.

Sometimes, though, the best way to enjoy a quiz is to test yourself, so the book is designed to be read through solo as well. Each answer is accompanied by illuminating information, so there is more to it than just getting the answer right. Of course, if you're using the book as a pub quiz, you don't need to include these parts.

If you are going to use the book in a quiz, you'll need to copy the questions from the two special rounds and print out enough so that each team can have their own question sheet. You might like to use one of these as a 'table' round, which is left on the teams' tables to answer between the other rounds.

A popular addition in quiz play is to allow each team to have a joker to use on a round of their choice (before they see the questions), which doubles their points in that round.

The little factoids after each question are primarily for your enjoyment, but depending on your audience, it might add to the fun to read them out when running a quiz. And if a topic takes your interest, each question has a 'Further reading' link to the book list at the back, to really delve into a subject.

However you use the book – enjoy it!

QUIZ 1
ROUND 1:
EARTH AND MOON

QUESTION 1
Counting moons

How many moons does the Earth have?

Answer overleaf

While you're thinking ...

Jupiter has at least 67 moons.

The largest moon in the solar system is Jupiter's moon
Ganymede, which has a radius of around 2,600 kilometres,
more than one third the size of the Earth.

There is evidence already of moons around planets in other solar systems.

The Earth has one moon

This may seem an obvious answer to a ridiculously easy question, but viewers of TV show QI have been told that it isn't true. While the show has been on air, the number they have provided has varied from 0 to 18,000 – but in reality, the obvious answer, 1, is the best.

The reason given for a large number is that lots of little lumps of rock get captured by Earth's gravitational field for a few days and while captured are natural satellites, making them moons. The zero figure suggests that the Moon is a planet, not a moon, because it is unusually large compared with the Earth – but this decision is arbitrary and is not accepted by the astronomical community. (And as 'the Moon' it is just a moon.)

There is not as definitive a definition of 'moon' as there is of 'planet', but there are still clearly intended consequences from using the word 'moon'. These are that the body in question should be:

- Long-lasting – I suggest staying in orbit for at least 1,000 years
- Sizeable – say at least 5 kilometres across

This would still allow moon status for the pretty dubious companions of Mars, Phobos and Deimos, which are about 20 kilometres and 10 kilometres across.

Clearly such rules are implied when we talk about moons. If the time rule didn't exist, then every meteor that spent a few seconds passing through our atmosphere would be a moon, while without the size rule, we would have to count every tiny piece of debris in Saturn's rings as a moon – each is, after all, a natural satellite.

*Further reading: **Near-Earth Objects***

Space Station blues

We've all seen astronauts floating around pretty much weightless on the International Space Station. What percentage of Earth normal is the gravity at the altitude of the ISS?

Answer overleaf ➜

While you're thinking ...

The first part of the International Space Station was launched in 1998.

The orbit of the ISS varies between 330km and 435km
above the Earth – call it 350km for this exercise.

One of the favourite sections of the ISS for astronaut photographs
is the Cupola, an observatory module that has been likened
to looking out of the Millennium Falcon in *Star Wars.*

At the ISS, gravity is around 90 per cent Earth normal

Allow yourself a mark for anything between 88 and 92 per cent. Newton gives us a value for the gravitational attraction (F) between two bodies as: $F = Gm_1m_2/r^2$.

We can use this to work out the difference between the ground and the ISS. Luckily, practically everything cancels out. G (gravitational constant) is the same, m_1 (the mass of the Earth) is the same and m_2 (the mass of a person) is the same. So the ratio of the gravitational forces $Force_{ISS}/Force_{Earth}$ is just r^2_{Earth}/r^2_{ISS}, where r_{Earth} is the distance from the Earth's centre to its surface and r_{ISS} the distance from the Earth's centre to the ISS.

We're saying the ISS is 350 kilometres up. And the radius of the Earth is around 6,370 kilometres. That makes r_{ISS} equal to r_{Earth}+350, or 6,720 kilometres. Not very different. So the ratio of the forces is (6,370 × 6,370)/(6,720 × 6,720) – which works out around 0.9. To be more precise, the force of gravity at 350km is 89.85 per cent of that on the Earth's surface.

So how come the astronauts float around, pretty much weightless? Because the ISS is free-falling under the force of gravity – which means it cancels out the gravitational pull. It might seem something of a headline news event that the Space Station is falling towards the Earth, but there's another part to the story. The ISS is also travelling sideways. So it keeps missing.

That's what an orbit is. The object falls towards Earth under the pull of gravity. But at the same time it is moving sideways at just the right speed to keep missing the Earth and stay at the same height. As a result every orbit has a specific velocity that a satellite needs to travel at to remain stable.

Further reading: **Gravity**

A question of dropping

Who dropped a hammer and a feather on the Moon to demonstrate that without air they fall at the same rate?
(For a bonus – which mission was it?)

Answer overleaf

While you're thinking ...

It is very unlikely that Galileo dropped balls of different weights off the Leaning Tower of Pisa to show they fall at the same rate. The story came from his assistant, shortly before Galileo's death. Galileo was a great self-publicist and would surely have mentioned it had it been true.

What Galileo did do, though, was compare the rate of fall of pendulum bobs and balls of different weights rolling down an inclined plane – much easier than getting the timing right with the Leaning Tower.

The Ancient Greeks thought that heavier objects fall faster because they have more matter in them, and matter has a natural tendency to want to be in the centre of the universe. So with more matter, a heavy object should have more urgency in its attempt to reach its preferred place.

David R. Scott dropped a hammer and feather on the Moon

I will let you off the middle initial – and have a bonus point if you knew that the mission was Apollo 15. Scott beautifully demonstrated that the only reason a feather falls more slowly on the Earth is because of the resistance of the atmosphere. (You can see him in action here: http://youtu.be/KDp1tiUsZw8)

The Ancient Greeks were perfectly capable of trying this out (not the hammer and the feather on the Moon, but dropping similar sized balls of different weights), but it didn't fit with their approach to science, which was all about logical argument rather than observation and experiment.

Although Galileo did plenty of experiments, which mostly confirmed that different weights fall at the same speed, he also found a logical argument that would have worked for the Greeks if they had thought of it, and that would have enabled a much earlier development of an understanding of gravity.

Galileo imagined you had two balls of different weights, and the heavier *did* fall faster than the lighter one. You would equally expect a third ball of the combined weight of the two to fall faster still. But let's make that third ball from two separate parts, one for each of the two original weights, joined by a piece of string. The heavier of the two should fall a bit more slowly than it would otherwise, because the lighter weight would slow it down. Similarly the lighter weight should fall a bit faster than it otherwise would. So, the connected weights should fall at an intermediate speed.

But that means the same weight, depending on whether or not it is split, has two totally different speeds – showing that the idea doesn't make sense.

*Further reading: **Gravity***

The black hole of Earth

If the Earth were made into a black hole, what would be the diameter of its event horizon?

Answer overleaf ➜

While you're thinking ...

It's often said that American physicist John Wheeler was the first to use the term 'black hole' in 1967, but it was casually in use at an American Association for the Advancement of Science meeting in January 1964, as a result of which it first appeared in print in a *Science News Letter* article by Ann Ewing. No one is sure who thought of it.

A black hole is a body that has been so compressed that gravitational attraction overcomes any opposing forces and it disappears to a point.

Although a black hole is a point with no dimensions, to the outside world it appears as a sphere known as the event horizon. This is the distance from the centre at which spacetime is so warped that nothing, not even light, can escape.

The event horizon of a black hole Earth would be 20mm

Allow yourself a point for anything between 15mm and 25mm and half a point for between 5mm and 50mm. The only known natural mechanism for black hole formation is when a dying star collapses, but in principle any chunk of matter could be converted into a black hole if it were sufficiently compressed, including the Earth.

Despite the Hollywood portrayal, a black hole is not a giant suction device that pulls in everything around it. Its gravitational attraction is just the same as that of the body that formed it – in the case of our hypothetical black hole, it would be the same as the Earth. But Newton made it clear that a gravitational body like the Earth acts as if all its mass were concentrated at its centre. The big difference between black hole Earth and the real thing is that you can get much closer to that centre of mass in the black hole version.

Since Newton's time we have known that gravitational force follows an inverse square law – it increases as the square of the distance between the centres of mass of the attracting bodies decreases. So, when the distance halves, the force quadruples. The radius of the Earth is around 6,370 kilometres, so by going from our usual position on the Earth's surface to that of the black hole's event horizon means that the distance has reduced by a factor of 637,000,000. Which results in the force going up by a factor of 405,769,000,000,000,000. That's a whole lot of gravitational attraction.

Further reading: **Gravity**

The man who fell through Earth

If you fell down an airless, frictionless hole going all the way through the Earth, how long would it take to fall to the other side? (To the nearest minute.)

Answer overleaf ➜

While you're thinking ...

The diameter of the Earth is approximately 12,470 kilometres.

Your maximum speed falling through the centre of the Earth would be around 7,900 metres per second.

When you arrived on the other side of the Earth you would have to be removed quickly, to avoid falling back through again.

It would take 42 minutes to fall through the Earth

Score one point for anything between 41 and 43 minutes. (Half a point between 37 and 47 minutes.) The calculation, which uses calculus to pull together the timing as the person accelerates towards the centre of the Earth and then decelerates on the way out, assumes, of course, that there is no air resistance or friction as you travel, so you would have to build an evacuated tunnel through the Earth to make this impressive free journey possible.

Building a tunnel through the centre of the Earth would be a major engineering challenge. Not only would you have to build a tunnel 100 times longer than anything now existing, you would have to cope with high temperatures, radioactivity, and protecting your passengers so that they can survive the journey. You would also have to suppress any resultant escape of molten materials and a general tendency to cause earthquakes and volcanic disruption that would not make you popular with those living near the tunnel.

Dropping through the tunnel, you would accelerate as you travelled, influenced by a gradually decreasing gravitational pull. Then at the centre, after a moment of zero gravity, you would start to slow down, coming to a stop just as you reached the far side.

Interestingly, the journey time of 42 minutes is not dependent on taking your tunnel through the centre of the Earth. If you go to one side, and so make a shorter tunnel, the acceleration will reduce accordingly, and the journey will still take 42 minutes. As long as you can make the tunnel airless and frictionless, the process should work.

Further reading: **Gravity**

Lunar currency

What sized coin, held at arm's length, would appear about the same size as the Moon?

Answer overleaf ➜

While you're thinking ...

One of the smallest coins minted in the UK is the silver
Maundy penny, donated to pensioners by the monarch
at the Maundy Ceremony – it is just 11mm across.

The biggest UK coin that might still fall within living memory
was the so-called 'cartwheel' penny, issued from 1797. It
was a massive 36mm across, but was no match for the
much rarer cartwheel two-penny piece at 41mm.

It is a pure coincidence that the Moon appears to be about
the same size as the Sun. The Sun is approximately 400 times
further away, but also around 400 times bigger.

No coin held at arm's length is small enough to match the Moon

There isn't a coin small enough – and there never has been. Remarkably, the Moon's apparent size is only about the same as the hole produced by a hole punch (around 5 millimetres), held at arm's length. You can test this out by holding a punched piece of card up to a full Moon – the whole thing is pretty much visible.

Why, then, does the Moon look so much bigger? It appears to be a psychological effect. Because the Moon is very bright in contrast to the dark night sky, our brains assume that it is bigger than it really is. The picture we see of the world is not like a photograph, which captures the relative placement of everything geometrically. Instead, the brain has a range of modules that handle aspects like shapes, shading, lines and so forth. The picture we 'see' is a construct rather than an actual photographic image. And this means that things really aren't always the way we see them. This is why good optical illusions are so convincing.

In the case of the Moon, it can seem even bigger when it is low in the sky, appearing relatively enormous when apparently near to buildings or trees. Again, the brain is misjudging the situation. Because this effect is subjective, we can't be certain that the Moon looks the same to other people as it does to us. And the lack of this effect is why photographs of the Moon taken with an ordinary camera look so disappointing. It takes a good telephoto lens to get the kind of big Moon our brains tell us is out there.

*Further reading: **Inflight Science***

Measuring the world

What's the distance from the North Pole to the equator through Paris to the nearest kilometre?

Answer overleaf ➜

While you're thinking ...

The Greek philosopher Eratosthenes was the first person known to make a scientific measurement of the circumference of the Earth, by measuring the Sun's position at noon in two well-separated locations and throwing in a spot of geometry.

The modern SI (Système International) unit metric system was established in 1960. The standard unit of length is the metre.

The US Congress ratified the use of metric units in 1866, and the conventional units used there, like the foot or pound, are defined from metric measures – but despite attempts to introduce it, the metric system has never been accepted by the American population.

It's 10,000 kilometres from the North Pole to Paris

One point for the exact value, half a point for 100 kilometres either way. This bizarrely accurate measurement reflects the definition of the metre in 1795 as 1/10,000,000th of the distance from the pole to the equator through Paris. This distance was used to set up a platinum metre bar in 1799. Variants on this were used through to 1960, when the standard was moved to a definition based on wavelengths of light. In 1983 it became 1/299,792,458th of the distance light travels in a second.

The metric system is now common throughout the world, though it has been less well adopted in English-speaking countries, possibly because of an aversion to anything of French origin. The metric system is based on that adopted in France after the Revolution during the 1790s. It was extended from the original length and weight measurements around 40 years later, and the SI system is now a comprehensive system of measurement covering everything from the brightness of light (candela) to the number of particles in a standard quantity of matter.

The name 'metre' originated well before metrication, in the 17th century, when it was suggested that there should be a 'universal measure' of distance, or in Italian a *metro cattolico*. The 'kilo' part, derived from the Greek term for a thousand, was added as an official standard in 1960, but had been used since French metrication.

*Further reading: **The Story of Measurement***

Eggs in space

Which international company sponsored an International Space Station experiment to see if quails' eggs would develop in microgravity?

Answer overleaf ➔

While you're thinking ...

A quail's egg is about 1/3 the length of a chicken's egg, making it ideal for the limited accommodation in a space station.

An experiment using eggs on the International Space Station was originally thought up by an eighth-grade student, John Vellinger.

To enable the experiment, NASA had to deploy the ADF and the AHH on the ISS. Who said TLAs (Three-Letter Acronyms) are dead?

KFC sponsored the ISS quails' eggs experiment

The microgravity experienced on the International Space Station (ISS) is ideal for experimenting with the impact of gravity (and its absence) on growing things. Experiments there, for instance, were used to see how plants grow without gravity to direct their roots (answer: badly). But the outcome was worse for birds.

In the experiment sponsored by Kentucky Fried Chicken, a set of 36 Japanese quails' eggs were incubated on the ISS to see how the lack of gravity would influence their development. Being a NASA experiment, the equipment used was given impressive-sounding initials: the ADF or Avian Development Facility and the AHH (Avian Hatching Habitat). Special egg holders were designed to minimise the vibration that the eggs would suffer during launch, to avoid them being scrambled before they could be tested.

It turned out that an essential requirement for an egg to hatch was for the yolk to be kept relatively near the shell by the pull of gravity. In space, the yolks floated in the middle of the white and did not develop properly. None of the eggs hatched satisfactorily, though there was some development. The quails' eggs were used rather than the more KFC-friendly chicken eggs because they have less mass, and take up less room, so are cheaper to get into space.

The student who thought up the experiment, John Vellinger, later founded the Space Hardware Optimization Technology company. NASA paired Vellinger up with Mark Deuser, an engineer from KFC, to develop the original experiment.

Further reading: **Gravity**

QUIZ 1
ROUND 2: MISCELLANY

QUESTION 1
Spaghetti science

What is spaghettification?

Answer overleaf ➜

While you're thinking ...

Spaghetti is the plural of *spaghetto*, the Italian for 'thin string'.

The earliest known UK reference to spaghetti dates back to 1845.

Most pasta is a form of noodle, made from
unleavened durum wheat flour.

Spaghettification is the scientific term for the fate of someone falling towards a black hole

You also get a point if you came up with a description of what happens to a person during spaghettification, as detailed below.

A black hole is a theoretical construct that emerged early on from Einstein's gravitational theory, General Relativity. It was a good few years later that acceptable, if indirect, evidence was found to suggest that black holes do exist in the universe. (Or something close to a black hole, as some physicists dispute whether they can properly form.)

A black hole is a star that has completely collapsed. Usually the reactions in a star fluff it up, preventing it from collapsing, but certain stars, at a late stage of their development, aren't able to resist the attractive force of gravity and they collapse, ultimately reaching a point-like object or singularity. What is often labelled as a black hole is its event horizon, the distance from it at which nothing, including light, can escape.

If you were dropping into a black hole feet first, you would discover that the gravitational pull on your feet was greater than that on your head because your feet are closer to the black hole's centre of mass. What initially would be an irritation would become an excruciatingly painful and irresistible force, stretching your body longer and longer until you became a long, pink, spaghetti-like structure.

This process is given the name spaghettification. (Who says that cosmologists and astrophysicists don't know how to have a laugh?)

Further reading: **Before the Big Bang**

QUESTION 2

Winning physics gold

What won Gustaf Dalén the Nobel Prize for Physics in 1912?

Answer overleaf

While you're thinking ...

The first Nobel Prize for Physics was awarded in 1901
to Wilhelm Röntgen for his work on X-rays.

Only one person (at the time of writing) has been awarded the
Nobel Prize in Physics twice – John Bardeen. He won in 1956
for the research work leading to the transistor and again in 1972
for his contribution to the BCS theory of superconductivity.

At the time of writing, the youngest Physics Prize-winner was
Lawrence Bragg, who was just 25 when he won the prize in 1915, jointly
with his father, William, for their work on X-ray crystallography.

Dalén won the Physics Nobel for his work on gas lighting in lighthouses

There is some debate over whether the Nobel Prize should be awarded for the development of technology, rather than true physics, and few cases more deserve a raised eyebrow than that of Swedish engineer Nils Gustaf Dalén.

To quote the Nobel Prize website:

> For centuries, lighthouses have made navigation safer. During the 19th century, acetylene became the standard light source. Gustaf Dalén developed a method to flash the light in short frequencies thereby reducing gas consumption. In 1907, he invented a regulating valve based on the difference in expansion for black and white metal rods. The 'solar valve' made it possible to extinguish the light during day-time.

That this should be a reason for winning the Nobel Prize seems bizarre in the extreme, especially as by 1912, when he won the prize, gas lighting was rapidly being replaced by electricity. (The first lighthouse built specifically with an electric light was Souter Lighthouse in Tyne and Wear, opened in 1871.) It's a bit like someone winning a Nobel now for inventing the floppy disk.

While it's never possible to second-guess the workings of the Nobel Prize committee, it's worth remembering that the prize is awarded by the Royal Swedish Academy of Sciences, and it's just possible that they felt it was about time they had a Swedish winner, whatever his contribution to physics.

Further reading: www.nobelprize.org

The Carrington question

What was the Carrington event?

Answer overleaf ➔

While you're thinking ...

Richard Carrington was a British amateur astronomer with independent means who produced a survey of circumpolar stars but spent most of his time observing the Sun, particularly sunspots, discovering the way that the Sun's surface rotation is not uniform.

Galileo published his 'Letter on Sunspots' in 1613, suggesting that sunspots, dark patches that appear on the Sun, were phenomena on the surface of the Sun, rather than tiny planets getting in the way of the sunlight as had previously been suggested.

In the same document, Galileo suggested that the Sun rotates about once a month. The actual figure is a little over 25 days, so Galileo's measurements were reasonably good.

The Carrington event was a massive electromagnetic pulse caused by a solar flare

The Carrington event is named after the British astronomer Richard Carrington, who specialised in observing the Sun. In 1859, Carrington recorded a massive solar flare. It produced several coronal mass ejections, where clouds of charged particles are blasted into space, which subsequently reached the Earth.

Even though there was far less dependence on electricity back then, and electronics did not exist, the impact was significant. Strong electrical currents were induced in the wiring of the early telegraph systems. Sparks flew, some telegraph offices were set on fire and whole systems were put out of action.

These natural solar electromagnetic pulses (EMPs) continue to happen on an irregular basis. There was a major EMP, for example, in 2012. As it happens, the Earth was not in the line of fire, but had it been, the chances are that all our satellites would have been permanently wrecked, most of our terrestrial communications would have been taken out and many electronic devices would have been ruined.

A study by the US National Academy of Sciences puts the financial impact of a strike by a significant solar EMP at over $2 trillion, with repairs to systems and technology taking years to complete. Luckily, direct hits are infrequent – we haven't had one since the Carrington event – but there is little doubt that one will come our way eventually.

*Further reading: **Ten Billion Tomorrows***

An absolute temperature

Why is the temperature 2.725 K significant?

Answer overleaf ➜

While you're thinking ...

When a temperature is shown with a K, e.g. 54.6 K, the 'K' stands for Kelvin. The unit is named after the 19th-century physicist William Thomson, who took the title Baron Kelvin.

The units are called kelvin (with a small k), and have no 'degree' attached to them, unlike Celsius or Fahrenheit, though the size of a kelvin is the same as that of a degree Celsius.

The Kelvin scale starts at absolute zero, which is −273.15°C, and is frequently used in scientific measurement.

2.725 K is the 'temperature' of the Cosmic Microwave Background radiation

Clearly this is very cold (no points for that, I'm afraid), just a couple of degrees above absolute zero. The temperature 2.725 K is the equivalent of –270.425°C or –454.765°F.

The Cosmic Microwave Background radiation is a faint microwave 'glow' that fills the skies from all directions and is thought to be the remnant of the photons that escaped when the universe first became transparent, a little under 400,000 years after the Big Bang. At the time these photons were high-energy, but as the universe has expanded they have been 'red shifted' all the way down to the microwave region.

This was predicted a couple of decades before it was (accidentally) detected, with the original estimate being around 5 K – not a bad guess. It seems odd that the radiation is usually referred to by its temperature, rather than the more direct measure of frequency or wavelength that is usually given to light. It reflects the fact that the radiation has a distribution of frequencies. Light doesn't really have a temperature – what the 2.725 K refers to is the temperature of a piece of material (technically a 'black body') that would emit radiation of this energy, which comes in a frequency range of around 3×10^8 to 3×10^{11} Hz.

Further reading: **Before the Big Bang**

QUESTION 5
The movie snack menace

How did James Vicary fool the world using pictures of Cola-Cola and popcorn in a movie theatre in 1957?

Answer overleaf

While you're thinking ...

James Vicary ran a market research company and has an interesting association with Coca-Cola.

Coca-Cola has long been one of the world's most powerful brands. It was produced as an alcohol-free version of its inventor Colonel John Pemberton's mix of wine and cocaine (a drink he seems to have devised to wean himself off morphine).

Originally a patent medicine, Coca-Cola was first sold in 1886. Cocaine was no longer added after 1903, though there was trace cocaine from the coca leaves used in the process. The drink is now made with a cocaine-free coca extract.

Vicary claimed to have influenced the audience's purchases using subliminal advertising

James Vicary claimed to have exposed over 450,000 people to subliminal images, shown for such a short time in a movie that they weren't aware of them. These images showed Coca-Cola and popcorn, and Vicary claimed that there was an 18.1 per cent increase in Coke sales and a 57.5 per cent increase in sales of popcorn. Subliminal advertising was made illegal in the UK and a number of other countries.

It now seems that Vicary's 'experiment' was a publicity stunt to gain visibility for his market research company. Neither he nor others were able to reproduce his results, but this hasn't stopped the infamy of the experiment making many wary of subliminal effects.

There is some evidence that subliminal messaging does work, but not to the extent Vicary suggested. What it seems to do best is priming, a mechanism by which the brain can be made more aware of a particular topic. So, for instance, if someone is already thirsty, then flashing up an appropriate drink-related image or word will increase the chances that they will go and buy a drink – and can make them more likely to want a particular brand.

Vicary even got this wrong. He claimed that his experiment showed that you could make people want a drink or to eat, but that you could not get them to switch brands.

*Further reading: **Ten Billion Tomorrows***

QUESTION 6
A distinguished ology

What is ethology?

Answer overleaf ➜

While you're thinking ...

British Telecom's TV adverts featuring Maureen Lipman have
remained surprisingly memorable since being broadcast
in the 1980s. In one, she recovers from hearing that her
grandson got only two exam passes in pottery and sociology
by saying: 'He gets an ology and he says he's failed ...'

The use of 'ology' as a word dates back at least to 1811. In *Hard Times*
(1854) Dickens wrote: 'If there is any Ology left ... that has not been
worn to rags in this house ... I hope I shall never hear its name.'

One of the first 'ologies' after theology was demonology,
which dates back to the late 16th century.

Ethology is the study of animal behaviour

The name derives from the combination of the Greek for 'character' plus the -ology ending. You can also have the mark if you say that it is the portrayal of character through gesture, the science of character, or a treatise on manners or morals, but the most common use now is the zoological one.

Ethology is now strongly linked to evolution, starting with Darwin's 1872 book *The Expression of the Emotions in Man and Animals*. Although the book's primary focus is human expression, it has a foundation based on animal behaviour.

It has really been since the 1950s that ethology has taken off as a discipline. Animal behaviourist Konrad Lorenz suggested that animals had instinctive behaviours, set off by various external triggers – for example a mating dance or sensory stimuli associated with the animal's mother. As the field grew, it gave more consideration to the social side of animal behaviour, looking not only at individuals but how groups behave, all the way through to the complex interlinking of 'superorganisms' like bees and ants where the whole group of insects acts as if it were a single entity.

The social aspects of animal behaviour are particularly interesting from an evolutionary viewpoint, as they can seem at first to be counter-productive for an individual and its genes. However, group behaviour can often result in better survival chances for the majority, and a lot of work has been put into studying why and when animals will effectively sacrifice themselves for the greater good.

Ethology has side branches dealing with, for instance, the specific behaviour of members of the crow family and of dogs, both species that have been shown to be capable of remarkable feats.

*Further reading: **30-Second Evolution/If Dogs Could Talk***

HOW MANY MOONS DOES THE EARTH HAVE?

Elephant extraordinary

What links the elephant Tusko at Oklahoma City Zoo and Timothy Leary?

Answer overleaf ➔

While you're thinking ...

Fully-grown Indian elephants can weigh 5,000 kilograms.

The Oklahoma City Zoo opened for business in 1904.

Tusko was an Indian elephant – despite strong similarities, this is a different genus from the African elephant. Indian elephants are *Elephas maximus indicus,* while African elephants are *Loxodonta africana.*

Tusko the elephant and Leary both took LSD

Both the late Harvard lecturer Timothy Leary, intellectual pin-up of the hippy counterculture, and the elephant Tusko were subjects of experiments using the psychedelic drug lysergic acid diethylamide, or LSD for short, though a significant difference is that Leary had a choice in the proceedings.

After experimenting with the natural hallucinogen psilocybin, Leary became enthusiastic about the potential benefits of LSD in psychiatric care. A hero to 1960s youth culture, Leary popularised Marshall McLuhan's phrase: 'Turn on, tune in and drop out.' Leary would be repeatedly arrested, while his influence on the youth of America caused then US President Richard Nixon to call him the most dangerous man in America.

Leary made repeated experiments with LSD on himself and others. Tusko, by comparison, had a one-off experience with the drug. In the kind of experiment that it's hard to imagine being justified today, researchers decided to try a dose of LSD on the elephant. The justification was to see if it would induce the condition known as musth in which male elephants become uncontrollably aggressive. Unfortunately, the researchers working at Oklahoma Zoo had no idea what was an appropriate dosage of this powerful drug for an elephant. On 3 August 1962 they gave Tusko a dose that was between 1,000 and 3,000 times that taken by humans – but an Indian elephant has less than 100 times human body mass. Tusko collapsed almost immediately and was dead in less than two hours.

*Further reading: **Elephants on Acid***

It's Santa time

When it's 9pm on Christmas Eve in London, what time is it at the North Pole?

Answer overleaf ➜

While you're thinking ...

The concept of times being coordinated between locations largely came in with the railways. Before then, most towns operated their own time.

Despite often being confused, Santa Claus and Father Christmas are visually different. Traditionally, Santa Claus wears a short, belted tunic in red, trimmed with white and a separate hat – his look was strongly influenced by Coca-Cola advertisements. Father Christmas wears a full-length hooded cloak, which can be green, but is most often a more muted red than that of Santa Claus.

There are two North Poles. The geographic North Pole, the one referred to in this question, is the point at the northern end of the Earth where the planet's axis of rotation meets the surface. The magnetic North Pole is where the north pole of a freely rotating magnet points – and at the magnetic North Pole, the magnet will point straight down. Confusingly, the Earth's magnetic North Pole is actually what is called a 'south pole' when referring to magnets. This is because magnet poles were named after the pole of the Earth they were attracted to – and opposite poles attract. The two North Poles are around 800 kilometres apart, and the magnetic pole currently moves around 60 kilometres a year towards Russia.

When it's 9pm in London, it's 9pm at the North Pole too

The time on the surface of the Earth is arbitrary. We could have the same fixed time at every location – but that would mean that 9pm could be early morning, mid-afternoon or the middle of the night, depending on your location. Instead, the Earth was divided up into segments with different time zones, mostly during the 19th century.

So, for instance, the USA has four mainland time zones. Each zone around the world is, in principle, a bit like an orange segment, accounting for 15 degrees of the Earth's longitude, taking up the full 360 degrees in 24 hours. In practice, though, the segments veer around to match various boundaries, or to attempt to avoid a small island being divided into two zones. (China, which should straddle around five zones, has opted to go for a single massive zone.)

Most time zones have nice, clear one-hour boundaries, but some seem determined to confuse matters, as there are a few zones with 30-minute or 45-minute differences.

Such man-made confusion is as nothing to the North Pole. Here, every time zone meets – in effect, the North Pole is in every time zone. Take a single step away from the pole and whichever direction you choose should put you in a different time. This clearly isn't practical, and so arbitrarily, the time in the polar region is set at GMT – the same as the time in London during the winter. Hence Santa Claus is able to set his watch by Big Ben.

Further reading: **Build Your Own Time Machine**

QUIZ 1
ROUND 3: MATHEMATICS

Pick an answer, any answer

In multiple-choice tests with four options, which answer is most likely to be right?
a) The first; b) The second;
c) The third; d) The fourth

Answer overleaf →

While you're thinking ...

We don't know when the first multiple-choice test
was given, but they date back at least to 1915.

Multiple-choice became more popular with automatic marking.
The first readers used the electrical conductivity of pencil lead.

At least two-thirds of students taking multiple-choice tests think
that they should stick to the first answer that comes to mind – and
even the majority of professors giving students advice say this.
However, over 60 studies on test results show that when students
change their answer after consideration they are more likely to change
from a wrong answer to a right one than from right to wrong.

b) The second answer is most likely to be correct

A good question-setter uses a random number generator to select the positions of the correct answers, but often test writers use a distribution of answers that looks random, and human beings find generating random patterns difficult.

For questions with four options there is a tendency to have more answers with the second option correct – and, as it happens, it's true here. It's not a huge bias, but if you are guessing, you might as well guess the answer that is most often correct. With five answers, it's the final option that scores slightly better, but there is no clear bias with three.

If the choice is between 'true' and 'false', there are almost always more 'true' answers. This also applies to blanket answers like 'all of the above' – when available, they are used more often than random chance allows.

We also assume that repeats are less likely in random values than they really are. So a question-setter may use the same position for a multiple-choice answer twice in a row less than they should. If you are confident of an answer, but haven't a clue about the next, a useful way of eliminating an option is to avoid repeating the answer number from the previous question.

Another tactic is to eliminate answers that stand out. Take, for instance:

a)   Norman Smith
b)   Norman Spencer
c)   Stanley Spencer
d)   Arthur Eddington

… in which you can probably eliminate d). And finally, longer answers are often preferable because the test writer will try to make the correct answer *clearly* correct, which may need extra information.

Further reading: **How to Predict the Unpredictable**

The football player's nightmare

What is a professional footballer's best tactic to successfully kick a penalty?

Answer overleaf ➜

While you're thinking ...

According to an analysis in 2012, England had the worst record of any major international team in terms of penalty shootouts won, with just 14 per cent successful.

By contrast, Germany had 83 per cent successful shootouts. Bear in mind, though, that these figures are based on small samples – typically just six or seven shootouts.

That poor England result doesn't mean they didn't score. In fact, they succeeded on 66 per cent of the penalty kicks – but that was fewer than most of their opposition.

When taking a penalty, a professional should kick straight at the goalkeeper

I am the last person to give advice on football, but there is good evidence that this is true.

A goalie does not have time to wait until he or she can detect the path of the ball before deciding which way to dive. The decision has to be made based on the actions of the penalty-taker leading up to the kick. If the dive is any later, the ball will be in the net before the goalkeeper is in its way. But this means that the goalie is open to an unexpected tactic from the player. The assumption is that few professional players would risk kicking straight at the goalkeeper. So the goalie will almost always dive one way or the other.

This means that a bold player who kicks straight at the middle of the net where the goalkeeper is standing will have a higher chance of scoring. It's not a tactic that can be repeated indefinitely, but as an occasional ploy it tends to be unbeatable. So why don't we see it more often? It seems likely that many players would rather fail to score than risk kicking straight at a goalie who doesn't dive, leaving the penalty-taker looking stupid. They could not live down such an embarrassment, so they don't take the risk, even though it's the best way to score.

You might have taken part in amateur games where people regularly kick the ball at the goalie. This is why the question specified 'a professional footballer'. An amateur goalkeeper is more likely to wait, trying to see where the ball is going. This means that they will let a lot more shots into the corner of the net, but will stop those 'down the line' shots.

Further reading: **Think Like a Freak**

An average kind of question

How can the vast majority of a group of people be below average for that group?

While you're thinking ...

In case your statistics is rusty, an average, or mean, is simply found by adding up the value for each member of the population and dividing the outcome by the number in the population.

'Average' as used in non-technical speech has strange connotations. It manages to imply both typical and sub-standard at the same time.

Michael Gove, who was UK Education Secretary in 2012, caused a furore when he said that he wanted all schools rated 'good', where 'good' meant that pupil performance exceeded the national average. What he actually meant is that he wanted all schools to be rated 'good' in the future, in comparison with the *current* national average – but it came across as demonstrating an ignorance of mathematics.

If one individual in a group is a long way from the mean, most are that side of average

(Or if a small number of people are.) Imagine a room containing the entire staff of a supermarket. We're going to work out their average net worth – the value of their assets. Shelf stackers may have considerably less than managers, but the chances are they're all in the –£50,000 to +£500,000 range. Now let's have Bill Gates walk in. At the time of writing, his net worth is about £53,000,000,000. That is so much more than the rest that everyone except Bill is below average.

With this kind of distribution, where you have a relatively low number of individuals that are very different from the rest of the population, a mean average is rarely a useful measure. Instead, statisticians often use the median, which is simply the middle value, if you imagine all the different values set out in a line.

So, for instance, the mean UK house price isn't very useful, as the vast majority of house prices are below average. Statisticians will use the median, which the press will simply report as the 'average' because they think 'median' is too technical a term. That's what happens, for instance, with earnings, where it is the median that is used by government statisticians.

Sometimes it's more confusing still. I was suspicious a couple of years ago when a newsreader announced that the average house price in the UK was £163,910. The chief economist of Nationwide Building Society got in touch to explain that though they call it an average, it isn't. Instead they stratify the data according to region, type of house and so forth and produce a messy, weighted figure that could arguably be said to be a *typical* value – but that certainly isn't the average.

Further reading: **Dice World**

QUESTION 4
A random percentage

Multiple-choice: if you choose an answer to this question at random, what is the chance you will be correct? a) 25%; b) 50%; c) 60%; d) 25%

Answer overleaf ➔

While you're thinking ...

The percentage sign % is probably visually compressed from cto, a contraction of the Latin 'per cento'.

In this question, 'at random' can be taken to imply that each option has equal probability.

Most people are very bad at making a random choice. They have a preferred value, which they use too frequently.

Either 'none of them' or 'the question doesn't make sense'

The answer 'none of them' is reasonable in that if you thought of answering a) or d), then your choice has a 50 per cent chance of being correct, so it's wrong. Equally, if you thought of answering b) or c), your choice has a 25 per cent chance of being correct, so it's wrong.

Alternatively you could argue that the question is phrased ambiguously – what, for instance, does 'correct' mean in this context? – and so the question doesn't make sense.

Some versions of this problem are phrased without the 'multiple-choice' at the start, in which case 0 per cent is also a valid answer, but here it is ruled out by being labelled a multiple-choice question.

Further reading: **My Best Mathematical and Logic Problems**

QUESTION 5

Robert Recorde's claim to fame

When Robert Recorde wrote in 1557, 'I will sette a paire of paralleles ... because no 2 things can be moare equalle', what was he using for the first time?

Answer overleaf ➔

While you're thinking ...

Recorde was a Welsh mathematician who was also a medical doctor.

Something of a popular mathematician, Recorde wrote books intended for a wide audience on arithmetic, algebra, geometry and astronomy.

Sadly, despite being controller of the Dublin Mint at one point, Recorde made a bad political enemy in the Earl of Pembroke, who sued him for defamation. Recorde was put into a debtors' prison, dying there in 1558.

Recorde was using the equals sign (=)

It is hard to imagine mathematics without the familiar symbols, but they weren't always used and some have a known origin. Recorde also introduced + and − into Britain, though both had been in use in Germany since the mid-15th century.

To see how messy life was without symbols, in Ancient Greece this was the only way to write what we would now put as A + B = C + D:

THEAANDTHEBTAKENTOGETHERISEQUALTOTHECAND
THEDTAKENTOGETHER

(Though in Greek.) It didn't help that they also didn't bother with the 'space' symbol between words.

Sometimes, like Recorde, the person who introduced the symbol explains why they did so, but others remain a mystery. Take, for instance, the lemniscate, which is the name for the 'figure eight on its side' symbol ∞ we use for infinity (or, more precisely, potential infinity). This was introduced by the mathematician John Wallis in the mid-17th century in a book on conic sections.

Imagining a plane being made up of an infinite number of extremely narrow horizontal rectangles (or to be exact, parallelograms), Wallis casually introduces the symbol:

> Let the altitude of each one of these [parallelograms] be an infinitely small part, 1/∞, of the whole altitude and let the symbol ∞ denote Infinity.

However, no one has a clue why he used this new (and hence irritating to printers) symbol, rather than just a letter.

*Further reading: **A Brief History of Infinity***

Irrationally diagonal

Why is the length of the diagonal of a square with sides 1 unit long called 'irrational'?

Answer overleaf →

While you're thinking ...

The length of the diagonal of a rectangle can be calculated from the length of its sides using Pythagoras' theorem.

One thing we know for certain about Pythagoras' theorem is that it was known (at least pragmatically) long before Pythagoras was born.

Pythagoras and his followers believed that the universe was built on a structure of numbers.

Because the length of the diagonal is not the ratio of two whole numbers

Remarkably, the Pythagoreans were able to prove that this was the case. (And as they believed the whole universe was built on whole numbers, this rather unnerved them.)

Here's the proof. You don't have to read it if mathematical proofs leave you cross-eyed, but it is surprisingly simple.

Let's assume there is a ratio of whole numbers that *does* make $\sqrt{2}$ (which Pythagoras tells us is the size of the diagonal). We'll call the smallest such numbers A and B. So $A/B = \sqrt{2}$.

Which is the same as $A^2/B^2 = 2$; or, multiplying both sides by B^2, it is $A^2 = 2 \times B^2$.

Now $2 \times B^2$ must be even (anything multiplied by 2 is even). So A^2 is even. So A must be even (because an odd number multiplied by itself is odd).

That means A can be divided by 2. So A^2 can be divided by 4. So B^2 can be divided by 2. Which means B^2 is even. So B must be even (because an odd number multiplied by itself is odd).

If A and B are both even, they *aren't* the smallest whole numbers whose ratio is $\sqrt{2}$ – meaning that it's not possible to have a ratio of whole numbers that gives the correct answer.

Further reading: **A Brief History of Infinity**

Seven-league science

If you took one step each second in seven-league boots, what would your speed be in miles per hour?

Answer overleaf ➜

While you're thinking ...

'Seven-league boots' turn up in a wide range of European folk tales, from the work of the Brothers Grimm through Goethe's *Faust* to 'Jack the Giant Killer'.

The term seems to have come from the French *bottes de sept lieues*, featuring in the tales of Charles Perrault, though the concept is probably older.

It's not entirely clear how wearers of seven-league boots progress, but for the purposes of this question, assume that they take a normally timed stride, which covers seven leagues, rather than floating across seven leagues at walking speed, which would not be any quicker than ordinary boots (if less tiring).

Your speed would be 75,600 miles per hour

A league was the distance walked in an hour – around the three-mile mark – so by taking a 21-mile pace every second you would travel at 75,600 miles per hour. (Have a point for anything from 70,000 to 80,000 miles per hour, or 100,000 to 130,000 kilometres per hour.)

It would be a little windy, to say the least, and as you would be moving at over 100 times the speed of sound, you would also produce an impressive sonic boom. But very few stories seem to have thought through the practical consequences, which in a way is rather strange, considering the worries that were expressed about people not being able to breathe when travelling at 30 miles per hour on a railway train when they were first introduced.

Magic being what it is, the suspicion has to be that the boots don't really move their wearer at this speed, but somehow tunnel from start to arrival point without moving through the space in between, in the manner of a wormhole in space. In fact, at least one modern story featuring seven-league boots (the highly entertaining *Bartimaeus* trilogy by Jonathan Stroud) does involve some kind of spacetime warp. Similarly, Terry Pratchett warns in his *Discworld* series that special magical preparations are required, as the outcome of having your feet 21 miles apart is otherwise rather distressing.

Further reading: **Ten Billion Tomorrows**

Coasting along

How far is it round the coastline of Britain?

Answer overleaf ➜

While you're thinking ...

For the purposes of this question, the coastline of Britain
incorporates mainland England, Scotland and Wales.

You can't get further than 115 kilometres (70 miles)
from the sea in Britain.

Cornwall is generally cited as the county with the longest
stretch of continuous coastline in Britain, though Kent and
Essex have also been known to make the claim.

It's impossible to say how long the coastline is

You can also score for 'there is no fixed value' or words to that effect. The arbitrary figures you see published vary from 2,800 to 18,500 kilometres. In reality, this genuinely is a measurement that doesn't have a meaningful answer, thanks to the wonder of fractals.

Fractals are chaotic geometric patterns that are described as 'self similar' – meaning that if you take a bit of the pattern it looks similar to the whole. Think, for instance, of a naturally crinkly thing like a fern or a range of mountains.

If you are set the task of measuring the coast, you could take a measuring wheel and run it along a coastal path, or you could imagine laying a very long piece of string along the path, then picking it up and measuring the string. And that would give you a figure within the range above, depending on how straight the paths were. So far, so good. But what if you didn't stick to the path, but instead ran the string into all the big crinkles and bumps that lie along the edge of the coast that the path doesn't bother to go into?

Then, inevitably, you would measure a longer distance. And if you took a bit more care, and folded the string into the smaller crinkles and bumps, it would be even longer. If you had thin enough string you could do this all the way down to the bumps in the atoms making up the rocks – and then you would measure something vastly longer than the original path. Which is the 'real' distance? Any and all of them.

Further reading: **Inflight Science**

HOW MANY MOONS DOES THE EARTH HAVE?

QUIZ 1
ROUND 4: BIOLOGY

QUESTION 1
Eye spy

What is the furthest you can see with the naked eye?

Answer overleaf ➔

While you're thinking ...

The human eye is very sensitive, able to detect just a handful of photons in the right conditions.

On a clear day, you can see across the narrowest part of the English Channel, which is 21 miles (33 kilometres).

The eye's light detectors are clumsily 'wired' back to front, a good example of evolution not always achieving the best 'design'. (Though recent research suggests that there are some benefits to this arrangement.)

The furthest you can see is around 2.5 million light years

Even locally we can see things a remarkably good distance away. Assuming good eyesight (as we will throughout) and a very dark night, you can see a candle flame around 10 miles (16 kilometres) away. Sadly, most of us live in places where street lighting and sky glow ensure that it isn't dark enough to achieve this feat. However, our capabilities on Earth are dwarfed when you consider what is possible looking out into space.

Even in a town we can usually see a fair collection of stars. The easiest constellation to recognise is Orion with its distinctive three-star belt. The middle star in that belt, Epsilon Orionis or Alnilam, is around 1,340 light years away. That's the distance light travels in 1,340 years, which given its speed of 186,000 miles (300,000 kilometres) a second is quite a distance. That's about 12,686,155,200,000,000 kilometres away.

Quite impressive, but that's nothing to your eyes. Assuming a properly dark night, well away from a town, anyone with good eyesight should be able to make out the little fuzzy patch that is the Andromeda galaxy (check an astronomy site, or the book below, to find it). That is our neighbouring galaxy, around 2.5 million light years away. That's 1,865 times as far as Alnilam. When the photons that your eyes detect left Andromeda, the human race did not exist.

When we ask, 'How far can you see?', it's easy to limit your vision to the Earth – but the mechanism of seeing is no different whether we're looking across the room or across the galaxy.

*Further reading: **The Universe Inside You***

QUESTION 2

Locked up in a cell

Why are biological cells called 'cells'?

Answer overleaf ➜

While you're thinking ...

Cells are the smallest living item that can
reproduce without outside interaction.

The wall of the cell, the membrane, is a complex organic material that
allows some substances to pass through but resists the passage of others.
Some types of cells have a stronger outer wall outside the membrane to
provide protection. These are most common in bacteria, plants and fungi.

Cells are broadly divided into prokaryotes (bacteria and archaea), which
don't have a nucleus, have relatively few internal structures and make
up single-celled organisms, and eukaryotes, which have a nucleus, tend
to be more complex and are more common in multiple-celled organisms.

Because they look like the cells that monks used to live in

In 1665, Robert Hooke published his book *Micrographia*, a study of very small things with the relatively newly developed microscope. Apart from providing shocking large-scale drawings of a flea, a louse and the compound eye of a fly, Hooke examined thin sections of cork. He described seeing an 'infinite company of tiny boxes' making up the cork's structure.

To Hooke, these regular rectangular shapes looked like the rows of cells in the dormitory of a monastery, so that's what he called them. We now know that biological cells come in many shapes and sizes, and are far more complex than a simple box structure, but they retain this cross-over name.

A similar process has happened in physics. When cells were understood better, biologists noted that some kinds of cell have a central blob containing the DNA, known as the 'nucleus'. When Ernest Rutherford wanted a name for the (relatively) heavy central part of the atom that was discovered in his experiments, he borrowed the name 'nucleus' from biology. So terms like 'nuclear power' have an indirect link to Hooke's bedrooms for monks.

*Further reading: **The Universe Inside You***

QUESTION 3
Blood-red poser

What is it about haemoglobin that causes blood to be red?

Answer overleaf ➜

While you're thinking ...

Haemoglobin is a kind of nickname, a shortening of the original name for this protein of 'haematoglobulin', a word that combines the Greek for blood and the idea of a globule – it's a little blood blob.

Apart from water, over 95 per cent of a red blood cell is haemoglobin.

Foetal haemoglobin has a different structure from the adult form, with a better ability to grab onto oxygen, necessary because oxygen is in relatively short supply in the womb.

The red colour comes from porphyrin, a collection of organic ring structures in haemoglobin

The popular explanation that blood gets its red colour because of the iron in haemoglobin is wrong. A haemoglobin molecule does contain four 'heme' units, each with a single iron atom at its heart. And it is true that iron oxide (most obviously in the form of rust) has an orangey-red colour. So it seems a fairly reasonable leap to assume that the iron in haemoglobin is responsible for blood being red. But iron compounds don't have the bright red of blood, which comes instead from the array of organic rings that surround each heme unit in haemoglobin.

In a sense, it shouldn't be too surprising that the colouration is not down to the iron, as the iron atoms form only a tiny part of the much larger structure. Porphyrin gets its name from the Greek word for a reddish-purple hue. The actual colour produced by the molecule depends on the shape of the porphyrin structures, a shape that changes when oxygen is bound to the iron, meaning that oxygenated blood has a brighter red colour.

A change of shape also occurs during carbon monoxide poisoning, which happens when carbon monoxide binds to the heme units in preference to oxygen, producing a red flush in the victim.

Further reading: **The Universe Inside You**

Avian altitude

To the nearest thousand feet, what is the highest a bird has been observed flying?

Answer overleaf ➜

While you're thinking ...

As feet are still the standard height measurement in aviation,
we are giving the metric system a miss for once.

Airliners typically cruise in the range of 30,000 to 42,000 feet.

Light aircraft typically operate at up to 12,000 feet.

The highest altitude a bird has been observed at is 38,000 feet

This was decidedly unusual. Most garden birds don't go above 2,000 feet, and the more strenuous water birds rarely exceed 4,000 feet – but there are some special cases. Among the ducks, mallards have been seen at around 20,000 feet, while whooper swans have made it to 27,000 feet, and bar-headed geese, the most high-flying of the migrating birds, can reach 30,000 feet, where they make use of the jet stream to travel at high speed with respect to the ground.

The world record holder, though, is a type of vulture, a Rüppell's griffon, which has a 3-metre wingspan and was discovered at 38,000 feet. Unfortunately, this discovery happened the hard way, when the bird was sucked into an aircraft engine, so it wasn't around to see its entry in *Guinness World Records*.

Other living things are in the high-flying club. Bumblebees are found much of the way up Everest, and can in principle fly as high as 30,000 feet, though they are unlikely to be found far from mountains. The true record holders, though, are bacteria, which are found in large quantities in mid-air.

Over some cities there are as many as 1,800 different species of bacteria present in the air, and being so light and easily buoyed up, they are found far above anything else living (except humans). The highest recorded discovery is at about 65,000 feet. For comparison, Concorde, the highest-flying airliner so far, cruised at 56,000 feet and did not exceed 60,000 feet.

Further reading: **Inflight Science**

Hairy problem

To the nearest thousand, how many hairs are there on a typical human head?

Answer overleaf ➜

While you're thinking ...

Human hairs are typically between 0.02 and 0.2 millimetres across.

Hairs are dead. Despite everything the shampoo
makers tell you, they can't be 'nourished'.

Human hair grows at around 10 to 15 millimetres a month.

There are about 10,000 hairs on a human head

Score a point for anything from 9,000 to 11,000. We don't, of course, all have the same number of hairs (and for bald people, it's more a matter of hair follicles than actual hair). If you are naturally blond, you are likely to have more hairs than average, and red-haired individuals will have somewhat fewer. (This continues to be true if, like me, you have 'red hair' that is now grey or white.)

The hair on our heads is usually more visible than body hair, but overall we have a similar number of hairs to those on a chimpanzee – it's just that our body hair is mostly very fine and small, making it difficult to see. It's because of this fineness that we can see goose bumps, which occur when our body attempts to fluff up our 'fur', either to make it more insulating or to make us look bigger when in danger. Unlike most mammals, this is ineffective in us because the hairs aren't obvious, so we just see the impact on the skin.

Exactly why the hair on our head is so different from body hair is not known for certain. Apart from being less obvious, body hair also differs in not growing further after reaching a roughly fixed length, whereas head hair grows indefinitely. It has been suggested that this might be a side-effect of wearing clothes (but not hats); a way to protect the skull from impact or the sun; or an unexpected side-effect of a mutation that proved to have another benefit.

*Further reading: **The Universe Inside You***

QUESTION 6
Lousy ancestry

Which came first, the body
louse or the head louse?

Answer overleaf ➔

While you're thinking ...

The two types of lice are distinct species, but one evolved from the other.

When Robert Hooke published his book on the use of microscopes
in 1665, featuring large fold-out illustrations, the most fascination
and horror was generated by the image of a louse.

Adult head lice are around 2.5 to 3 millimetres in length.

The head louse came first

In the UK, at least, head lice are probably the most frequently experienced human parasites. There are regular outbreaks, especially among children at junior schools who are more likely to have the head-to-head contact required to pass on these tiny, blood-sucking insects.

Until somewhere between 50,000 and 100,000 years ago, the body louse did not exist. The head louse is quite fussy about staying near the bases of head hairs for security. But for some reason, back in the past, a variant developed that was prepared to risk the wide-open spaces of the body. We can estimate that this is the correct period of time from the variations in the DNA between the two species.

That timing gives one possible suggestion as to why these variants on head lice were suddenly willing to move away from the protection of the head hair. This also seems to be the time period in which humans started to move away from Africa, which itself could have been one of the triggers for starting to wear clothes. It seems that clothing not only kept these early humans warm but also provided a friendly environment for the most daring head lice, which evolved into their body lice cousins.

Further reading: **The Universe Inside You**

QUESTION 7
A rare question

What is the red liquid that oozes out of a joint of beef or rare steak called?

Answer overleaf ➜

While you're thinking ...

The best cuts of beef – silverside and topside
– come from a cow's rear end.

An increasingly popular cut of steak is the flat iron. (This is the American name – it was traditionally called butlers' steak in the UK.) Taken from the animal's shoulder, it is cut with the grain, unlike most steaks, so can be a little tougher, which means it can be bought at a better price.

Cooked red meat goes brown because iron atoms in compounds in the meat lose an electron to reach the ferric state. Meats processed with nitrites like ham tend to stay pink when cooked because nitric oxide becomes bound to the iron atom, preventing it from reaching the ferric state.

The red liquid is myoglobin (in water)

Some who like their beef (or steaks) well done, uniformly brown, tend to moan when another diner goes for a rare cut, disliking the way that it 'oozes blood'. But that red liquid – the same that gives raw meat its colouration and that drips from a piece of meat – is not blood.

When you think about it, what you see is quite different from the blood that emerges from a cut in your finger. Blood is opaque and strikingly red. The liquid coming out of meat is transparent, and a duller shade of red. It's not even dilute blood, but a totally different substance, myoglobin, suspended in water.

Like the haemoglobin in blood, myoglobin has the job of latching on to oxygen, but it is significantly less efficient. Each molecule of mammalian haemoglobin can carry four oxygen molecules, but myoglobin holds only a single oxygen molecule.

The functional role of the compound is different too. Haemoglobin is in the business of transporting oxygen around the body to where it is needed. Myoglobin acts as a temporary store for that oxygen, typically in a muscle, before it's put to use. Because they need to hang on to more oxygen than a normal mammal, natural divers like whales and seals have especially high concentrations of myoglobin in their muscles, enabling them to continue doing work for longer between breaths.

Further reading: **The Universe Inside You**

QUESTION 8
Arachnid anxiety

Why could there never be giant spiders (as in *The Lord of the Rings* or *Harry Potter*)?

Answer overleaf

While you're thinking ...

The world's largest known spiders are male goliath bird-eating spiders, which can have a leg span of up to 28cm, and giant huntsman spiders, which can reach around 30cm.

Spiders are, of course, not insects, but eight-legged arachnids, related to scorpions.

Most, but not all, spiders produce spider silk, which has an extremely high tensile strength for its weight, coming between steel and Kevlar.

Because they would collapse under their own weight – and would also asphyxiate

Have a point for either – and a bonus half-point if you got both.

Few people like spiders – and they're bad enough when they're just a few centimetres across. Imagine one big enough for humans to be its prey. This would, perhaps, require magnifying the spider by 100 times. So its legs would be 100 times longer, its body 100 times as far across and so on.

If we did this, its legs would have an increase in cross-sectional area of $100 \times 100 = 10,000$ times, making them much stronger. Unfortunately, though, the spider's mass, which depends on its volume, would go up by $100 \times 100 \times 100$ times – making it a million times heavier. The result is that its legs would snap under its own weight.

This same effect makes it impossible to blow up humans vastly bigger than their current size. (If you take a look at a really large animal, like an elephant, it will have proportionately chunkier legs than you do.) However, spiders and insects have another problem. They breathe through tiny inlets in their carapace, the hard outer shell they have in place of a backbone. So a million times as much body would get only 10,000 times as much oxygen – if you could magically blow up a spider, it would rapidly asphyxiate.

Australians will tell you it's sensible to be wary of spiders in some parts of the world – but we don't need to worry about mutant giant spiders taking over.

*Further reading: **The Universe Inside You***

QUIZ 1
ROUND 5:
TECHNOLOGY

The wind-up king

Who invented the gramophone?

Answer overleaf →

While you're thinking ...

The first cylinder recordings were made on a tinfoil sheet.

The tinfoil was improved on by moving to
wax-coated cardboard cylinders.

Disc recordings were also initially made on wax, but this was
used as a resist (a substance used to block a corrosive material)
when etching the recording into a zinc disc. Once well established,
78rpm records were mostly made of shellac and rock dust.

Emile Berliner invented the gramophone

It's a common pub quiz error to give the answer of Thomas Edison. Edison invented the cylinder-based phonograph in 1877, not the first sound-recording device, but the first to be able to both record and play back. But the disc-playing 'gramophone' was the work of the German-born American inventor Berliner.

Berliner patented his invention in 1887 and within ten years it was not just well established in the US, but had spread to the UK in the form of 'The Gramophone Company', which would eventually become part of EMI. The name was still used on recordings through to the 1970s. Although there was no particular sound quality benefit of discs over cylinders, they had two huge practical advantages. One was storage – like scrolls, cylinders were clumsy to store in any number – and the other was production, as a disc could be stamped, where a cylinder had to be cut, a slower and more expensive process.

The original gramophone records were metal, but by the start of the 20th century they were typically made of the brittle lacquer shellac with a filler that was usually powdered stone. Over time, they were gradually replaced by the more flexible and robust vinyl, as costs for its production came down. This also resulted in a move from the old 78 revolutions-per-minute standard to 33⅓ rpm, 'long playing', microgroove vinyl records, with intermediate 45 rpm speed for the smaller singles. The slower 33⅓ rpm speed seems to have been developed to match the timing required when discs were used to accompany sound movies, while 45 rpm was calculated as an optimal speed for a desired level of sound quality.

*Further reading: **The History of Music Production***

HOW MANY MOONS DOES THE EARTH HAVE?

Conscious computing

What do the initials HAL for the HAL 9000 computer mean in the film *2001: A Space Odyssey?*

Answer overleaf →

While you're thinking ...

The movie *2001: A Space Odyssey* (1968) still holds up very well visually, despite being shot without CGI. For instance, the scenes in the *Discovery One* spacecraft, where the astronauts walk around the inside of a rotating wheel-like section, were filmed on a huge rotating set at Borehamwood Studios in London.

When *2001* was released, the year 2001 seemed a long way ahead. As usual, reality moved on further in some ways (no Pan Am, for instance, a strongly featured brand in the movie), and lagged behind in others – we have no moon base, nor conscious AI computers like HAL.

The on-board computer HAL was scripted to show more emotion than the human space travellers in *2001*.

They are an acronym for 'Heuristically programmed ALgorithmic computer'

Arthur C. Clarke and Stanley Kubrick firmly denied the rumour that HAL was dreamed up by shifting each letter of IBM one place backwards in the alphabet. If you said it came from IBM, you don't get a point – but I won't take a penalty point off, as it did seem that Clarke and Kubrick protested a little too strongly.

When interviewed on the subject, Clarke said: 'I was embarrassed by the whole affair, and I felt that IBM, which was very helpful to Stanley Kubrick during the making of 2001, would be annoyed.' He went on to agree that it seems an unlikely coincidence as there was a 17,576 to 1 against chance of coming up with these initials accidentally. And it didn't help that IBM also had number-based series, so identifying HAL as a HAL 9000 series computer fit very closely.

In practice, though, Clarke was a little hard on himself. Assuming the computer was going to be given a name-based acronym, which was likely for such a device, there are only a handful of three-letter choices (three letters probably was because of IBM), like JIM, MAC, SAL, SAM and TOM. This makes the coincidence less surprising. Devising an easy-to-use acronym first and then deciding what the initials meant was fairly common practice in the IT world in the 1960s and 70s.

Further reading: **Ten Billion Tomorrows**

A light-bulb moment

Who invented the traditional incandescent light bulb?

Answer overleaf ➡

While you're thinking ...

Humphry Davy used the generator invented by his one-time protégé Michael Faraday to heat a platinum wire until it glowed, but the wire never survived long enough to be used for lighting.

The electric arc light was deployed commercially from the 1860s, over fifteen years before the incandescent light, but was too dangerous for domestic use.

The 2014 Nobel Prize in Physics was awarded to Isamu Akasaki, Hiroshi Amano and Shuji Nakamura for the invention of efficient blue light-emitting diodes, which made it possible for low-energy LEDs to replace older light bulbs.

Sir Joseph Wilson Swan invented the light bulb

I will allow you half a mark for Thomas Edison, but the British inventor Swan demonstrated an incandescent bulb that used a carbon filament just as Edison's did, in the same year as Edison, 1879, but eight months earlier.

Swan was more of a scientist than Edison, who had a business-oriented focus on invention for profit, and Swan had not applied for the same range of patents as his American rival. The 'Wizard of Menlo Park', as Edison was known, promptly sued for patent infringement.

In most cases, that would have been the end for Swan's claim to the invention, but the court recognised that Swan had been there first and threw out Edison's claim. Worse, as far as Edison was concerned, he was obliged to recognise Swan's priority by setting up a joint company, the Edison and Swan United Electric Light Company, to exploit the invention.

It's popular these days to belittle Edison and to claim that practically everything he was known for was either the work of Tesla or other engineers in Edison's massive invention factory. But this is unfair: Edison certainly was hugely inventive and did come up with a wide range of products, including the electric light bulb. He just happened not to be the first in this case.

Further reading: Light Years

The Big Blue blues

Around 1943, how many computers was IBM's Thomas J. Watson said to have predicted the world would require?

Answer overleaf ➔

While you're thinking ...

IBM was formed from the Computing Tabulating Recording Company, which made machines that tabulated punched cards. Nicknamed 'Big Blue', it became by far the biggest computer company in the world for many years.

Although it is arguable that IBM never really understood personal computing, the introduction of the IBM PC indubitably moved personal computing from a hobby to something for everyday homes and businesses – and made Microsoft what it is today.

Thomas J. Watson, Senior was at the helm of IBM and its predecessor from 1914 to 1956. Watson's motto, 'THINK', was used throughout IBM to encourage the staff to bring new ideas to the way they worked.

Watson predicted the world would need five computers

The rather clumsy wording of the question reflects the doubt that now exists over whether Watson ever made the remark. It seems to have been referenced only from the 1980s (though, to be fair, until then it seemed relatively unremarkable), and it is entirely possible that Watson never said it.

However, there was a widespread feeling in the 1940s that computers would always remain a specialist device that would never need to be widespread – and with the technology getting faster all the time, it seemed doubtful that they would ever need to be common. So even if Watson never actually said it, the myth reflects the sentiment of the times.

Now, of course, many of us have more computers than this in a single house, just counting laptops, desktops, smartphones and tablets. Once you bring in the computing power that is built into everything from a car to a washing machine, the numbers spiral upwards.

It is almost impossible to come up with a realistic number. For instance, how do you keep track of computers that have been put to one side after an upgrade, or scrapped? The number of desktops was estimated to pass 1 billion in 2002 and now has figures in the 1.5–2 billion range, while smartphones will probably overtake them around the time this book is first published, and tablets are already heading for half a billion. But are all these machines in use? What about the embedded processors we've already mentioned? All we can say for sure is it's a lot more than five.

Further reading: **When Computing Got Personal**

Atomic company

How many atoms did a team of IBM scientists use to spell out the letters 'IBM' in 1989?

Answer overleaf ➜

While you're thinking ...

'IBM' originally stood for 'International Business Machines',
but now the company is simply called IBM.

At the start of the 20th century most scientists believed
that atoms didn't exist, but were just useful concepts. It
wasn't until 1912 that French physicist Jean Perrin provided
experimental evidence that they were really there.

In 1980, Hans Dehmelt of the University of Washington
made a single ion (an atom with an electric charge) visible
with laser light, showing up as a tiny spot of brilliance where
the ion was floating in space, held in a magnetic trap.

They used 35 atoms to spell 'IBM'

One point for 35, half a point for anything between 30 and 40.

The original electron microscope worked by scanning a piece of material with a beam of electrons in a manner similar to a conventional microscope using light, but a later development known as the scanning tunnelling microscope made it possible not only to detect objects as small as a single atom but to manipulate them as well.

At IBM's research lab in Zurich in 1989, a team led by Don Eigler managed to arrange the 35 atoms of the inert gas xenon (which has the advantage of having relatively large atoms with atomic number 54) on a super-cooled sheet of crystalline nickel (this is often described as copper, but the original paper specifies nickel). The letters were five atoms deep so, for instance, the 'I' consisted of a five-atom downstroke with three-atom crossbars at the top and bottom.

The scanning tunnelling microscope uses a tiny tungsten probe, which creates a very small electrical potential difference between the surface and the tip of the probe. As the probe is scanned over the surface, it moves up and down, keeping the potential difference the same, which means that it 'feels' out the surface without ever touching it.

It was discovered by accident that at the right distance, the probe would attract an atom, making it possible to drag it across the surface, in this case to spell out the company's initials.

*Further reading: **The Universe Inside You***

Fizzing with ideas

Who invented man-made fizzy drinks?

Answer overleaf ➜

While you're thinking ...

The 'fizz' in fizzy drinks is carbon dioxide, hence the other common name, 'carbonated drinks'.

The earliest known water bottling plant in the UK is at the Holy Well in Malvern, Worcestershire, dating back to the 1620s.

Some spring water is naturally carbonated, though for practical reasons, the fizz is allowed to be removed and then re-added during the bottling process.

Joseph Priestley invented fizzy drinks

Priestley is best known as the discoverer of oxygen, but he also invented the manufactured carbonated drink a good few years earlier. This was in 1767, only eleven years after carbon dioxide had first been identified. Priestley was undertaking experiments at the Jacques brewery in Leeds (until he was thrown out for dropping ether into a brewing vat).

Specifically, Priestley was interested in the gas that was produced by the vats. Capturing some, he discovered that when it was bubbled through water it made the ordinary well variety taste like the exotic water from the Alps that contained natural bubbles and a zesty flavour.

Priestley seems not to have thought of this much further until he attended a dinner at the Duke of Northumberland's house. The Duke thought that science could be entertaining for his guests – hence Priestley's presence. Although he was soon to become the librarian for Lord Shelburne at Bowood House in Wiltshire, he didn't usually move in these circles. Northumberland served the recently developed, guaranteed-pure distilled water to his guests – but it seemed bland and unappetising. Priestley told the gathering he could make it palatable and came back next day with home-brewed soda water.

Though there is no doubt that Priestley got there first, the Swiss watchmaker Johann Schweppe was the first to produce soda water commercially, starting the massive industry we have today.

Further reading: **The Universe Inside You**

Patent madness

In 1930 Albert Einstein and a colleague were issued US patent 1781541. What was it for?

Answer overleaf ➔

While you're thinking ...

The co-inventor was Leo Szilard, a former student of Einstein's and a significant physicist in his own right.

The US patent office does not accept patent applications for perpetual motion machines unless they are accompanied by a working model.

Pioneer moving picture photographer Eadweard Muybridge held a patent for a washing machine.

Their patent was for a refrigerator

Fridges seem to work despite the second law of thermodynamics, which says that heat always flows from a hotter to a colder body in a closed system. The get-out clause is 'in a closed system'. A refrigerator gets away with it by consuming energy in order to pump heat to the outside, where it emerges from the radiator, usually located on the back.

When Einstein and Szilard invented their fridge, most current refrigerators operated using a poisonous gas as the refrigerant, which was kept under high pressure – an accident waiting to happen. In the 1920s, a German family was killed when the seal broke on the refrigerator and the gas escaped.

This led Einstein and Szilard to come up with a fridge design where there were no moving parts and the coolant was kept under constant pressure. It used a mix of two compounds, one of which could be quickly extracted to drop the pressure and hence the temperature. Einstein is usually thought of as unworldly and impractical, but he did have considerable experience of inventions from when he worked in the Swiss Patent Office.

This design is not widely used, but it has the advantage for developing-world applications of working with any source of heat, not just electricity.

Further reading: **What If Einstein Was Wrong?**

Matrix mechanics

In which year did the virtual reality world the 'Matrix', in which people could experience a lifelike, computer-based world – and would die in real life if they were killed – first reach our screens?

Answer overleaf

While you're thinking ...

When Neo is following instructions on his wonderfully dated-looking mobile phone in the first of the *Matrix* movies, we get one of the first examples of someone being told to take the door on the left, making a mistake and being told 'No! The other left!' (If anyone knows an earlier example of this usage, please drop me an email at brian@brianclegg.net)

The basic premise of the *Matrix* movies, that humans are being kept as energy sources, was terrible science – it's a very inefficient way to turn food into energy – but the movie was still highly entertaining.

The term 'cyberspace', for the computer-based world, dates back to 1982 when William Gibson used it in a story in *Omni*, two years before his game-changing novel *Neuromancer*.

The 'Matrix' first appeared in 1976

I'm afraid you get nothing if you said 1999, apart from a sense of being hard done by. Although the movie *The Matrix*, first of the trilogy, was released in that year, there was an earlier Matrix. This was in a 1976 *Doctor Who* serial called 'The Deadly Assassin'.

In this Tom Baker story, the Doctor visits Gallifrey. It's here that we come cross the Matrix, a huge neural network that enables participants to enter a virtual reality. The Doctor enters the Matrix to hunt down the Master, having been warned that should he die in the virtual world, he will die in reality as well.

I think it's unlikely that the Wachowski siblings (formerly Larry and Andy, now Lana and Andy), who wrote and directed *The Matrix*, were influenced by the *Doctor Who* episode, although they would have been ten and twelve when it was broadcast, and *Doctor Who* did have a strong cult following in the US. The underlying concept was widely used in science fiction, and while 'the Matrix' is not the only possible name for such an environment, it may well come in the top ten.

Further reading: **Ten Billion Tomorrows**

QUIZ 1
ROUND 6: CHEMISTRY

Molecular mindgames

What is the largest molecule that forms part of a human body?

Answer overleaf ➜

While you're thinking ...

A molecule is just a group of two or more atoms, which are linked by chemical bonds and where the group does not have an overall charge.

The human body contains around 7×10^{27} atoms, where 10^{27} is 1 followed by 27 zeroes.

Like all living things, the human body contains a large number of organic molecules, which can have complex structures and contain carbon.

Our largest molecule is chromosome 1

Half a point if you said DNA, which is what chromosome 1 is made from. Every normal cell in our bodies contains a pair of sets of 23 chromosomes. (A few specialist cells, like red blood cells, don't.) Each chromosome is a single molecule of DNA – something that isn't obvious when we see photographs of chromosomes, because the extremely long molecule is wrapped up around a kind of spindle, making it look much shorter and bulkier than it really is.

The DNA that makes up human chromosome 1 forms the largest molecule that is part of the human body, containing around 10 billion atoms. It is by no means the largest chromosome in nature – wheat, for instance, has a chromosome that is around four times as long – but it's our biggest.

DNA itself is described as a double helix, which refers to the outer structure of the molecule. This is made up of long sections (polymers) of sugars, which form the molecular backbone. Linking these strands at regular intervals are the important bits, the base pairs. These are joined pairs of relatively small organic molecules: cytosine (C), guanine (G), adenine (A) and thymine (T). The pairs always join the same way, C to G and A to T. This means that when a cell reproduces by dividing, the DNA can split into two parts, each of which contains all the information needed to recreate the other half.

DNA is, in effect, a family of molecules that act as data stores. Each base pair is the equivalent of a bit in a computer (though holding more information) and sequences of these biological bits, known as genes, act as the control codes for biological factories that produce essential molecules called proteins.

Further reading: **The Universe Inside You**

HOW MANY MOONS DOES THE EARTH HAVE?

QUESTION 2
It's a gas

How old is the hydrogen in your body?

Answer overleaf ➔

While you're thinking ...

Hydrogen is the lightest element, each atom
comprising a single proton and electron.

Typically 55 to 65 per cent of your body by weight is composed of
water, each water molecule with two hydrogen atoms. That's a whole
lot of hydrogen (not to mention all the hydrogen in organic molecules).

There is more hydrogen in the universe than any other element.

The hydrogen in your body is around 13.7 billion years old (probably)

One of my favourite exercises when talking to primary school children is to ask them how old they are. We start with the relatively small-scale biological possibilities, but then start to think about where the atoms in their bodies have come from. Clearly, initially they came from food and drink – but where before that?

There is always considerable excitement at the thought that atoms in their bodies have been in rabbits and foxes, in kings and queens, in trees and dinosaurs. But then the timescale gets pushed way back. The atoms already existed, I point out, when the solar system formed around 4.5 billion years ago. The heavier atoms will have been produced by stars and spread across the galaxy by explosions more than 5 billion years ago. But the really interesting one is the hydrogen. Because apart from a relatively few protons (hydrogen nuclei) produced by nuclear reactions, all the hydrogen in our bodies was formed shortly after the Big Bang, around 13.7 billion years ago. And it has been around ever since.

I put the 'probably' up in the answer (not required to get the point) to emphasise that while the Big Bang is our current best theory for the origins of the universe, it is by no means the only theory, and tests for theories in cosmology are inevitably significantly weaker than in other aspects of science. Along the way, the Big Bang theory has been patched up to match new data several times. It still holds up well, but cosmology has to be relatively speculative because of the indirect nature of the observations and the inability to undertake and replicate experiments.

*Further reading: **Before the Big Bang***

Bonding rituals

What temperature would water boil at if hydrogen bonding didn't exist?

Answer overleaf ➜

While you're thinking ...

Hydrogen bonding is an attraction between the relatively
positively-charged hydrogen in one molecule of water and
the relatively negatively-charged oxygen in another.

Hydrogen bonding is responsible for the way bonds are
twisted when water solidifies, pulling the molecules further
apart than at water's densest liquid form, which is why solid
water (ice) is less dense than liquid water, and floats.

It is sometimes said that water is the only substance that is
less dense as a solid than as a liquid, but this isn't true. It is
also the case, for instance, with acetic acid and silicon.

Water would boil at −100°C

The value is extrapolated, and usually given as either 'lower than −70°C' (which is true, if a little cautious) or −100°C, which relies on a straight-line extrapolation from other, similar compounds that don't undergo hydrogen bonding such as hydrogen sulfide and hydrogen selenide. Score a point for anything between −70°C and −110°C.

The implications are striking. If it weren't for hydrogen bonding, there would be no liquid water on the Earth. And all our experience suggests that while life can exist, for instance, without oxygen or at extremes of temperature that would kill a human being, all the forms of life we have so far come across require liquid water to survive. With no liquid water, Earth (and most otherwise inhabitable planets) would be unable to sustain life.

In fact, without the hydrogen bonding, water would not be so effective in supporting life anyway, even if it could exist at comfortable temperatures. It is as a result of the hydrogen bonding that water is a good solvent that is effective at transporting materials in cells. So these bonds are doubly important for the existence of life.

Further reading: **The Universe Inside You**

QUESTION 4
Helping the medicine go down

How many grams of sugar are there in a 200ml glass of skimmed milk?

Answer overleaf ➜

While you're thinking ...

There are 4–5 grams of sugar in a teaspoon.

A typical canned fizzy drink contains around 35 grams of sugar.

World Health Organization recommendations suggest
a limit on sugar consumption of 25g a day, up to a
maximum of 50g (regardless of gender).

There are 10 grams of sugar in a glass of skimmed milk

Give yourself a point for 1 gram either side of 10g.

Sugar is, without doubt, one of the least healthy contents of our diets and one most of us could do with reducing. In a way, cutting out fizzy drinks and sweets is the easy aspect of managing sugar intake. It's much harder when it comes to 'hidden' sugars. Sometimes these are the fault of manufacturers, who tend to load sugar into cereals, soup, baked beans and even bread to make them more appealing. But sometimes they are sneaked in by nature.

We know that skimmed milk is healthy milk. Like all milk it has useful calcium, famously good for teeth and bones. And it has a relatively tiny amount of fat – so simply switching from full-fat milk to skimmed milk is an easy way to reduce fat in the diet. But because it has that 'healthy' label, we don't consider the other things it might contain. But take a look at its dietary breakdown and you will find that skimmed milk has 5 grams of sugar for each 100ml of milk – surprisingly high in natural sugars.

Fruit smoothies are looked on significantly more dubiously by health experts – they pack in a startling amount of sugar from the fruit content, typically more than an equivalently sized fizzy drink. (Also, liquidising the fruit removes the dietary benefits of fibre, while the way its structures are destroyed makes it easier for the sugar to be absorbed.)

All fruits have some sugar content (which is why your 5+ a day should have more vegetables than fruit), but bananas, mangoes and grapes are among the worst, with around three times as much sugar by weight as cranberries or raspberries.

*Further reading: **Science for Life***

Water, water everywhere

How much water is there on Earth per human being?

Answer overleaf ➜

While you're thinking ...

According to environmental journalist Suzanne Goldenberg,
'The Middle East, north Africa and south Asia are all projected to
experience water shortages over the coming years because of decades
of bad management and overuse.' (*Guardian*, 9 February 2014)

Water shortages are predicted to 'pose threat of terror and war'.

Life and water are so tightly linked that H_2O is often used as a first
indicator that a planet or moon may be able to support life.

There are 210,000,000,000 litres of water per person

Give yourself a point for anything between 200 billion and 220 billion litres. That's a lot of water. If stacked up in litre containers, your personal pile would be around 10 million kilometres high. Let's assume you use around 5 litres a day. That would mean your water would last you for 115 million years. And that's assuming that you actually consume water – in practice almost all the water that goes into your body comes out again fairly swiftly.

Yet the people talking about increasing shortages of water in parts of the world are serious. An apparent part of the problem is that we don't use just 5 litres per person per day, but far more than that indirectly. Producing the meat for a single hamburger, for instance, can take around 3,000 litres of water. But once again that isn't water that disappears from the system. The real issues are twofold: water being in the wrong place and not being pure.

One impact of climate change is that some parts of the world are getting wetter and some drier. The dry bits are suffering increasingly because of this. Yet any nation with a coast is aware that there is plenty of water out there – it's just that it's salt water that we can't drink.

Both of these problems can be overcome. Water can be moved from place to place, and seawater can be desalinated. But these are expensive technologies – specifically, they are energy-intensive. Once the infrastructure is in place, the real issue is a shortage of cheap energy, not a shortage of water. Which is why, arguably, we should be putting far more effort into ensuring we have the best ways of generating energy.

*Further reading: **Inflight Science***

QUESTION 6
Popping pills

Why is 'aspirin' not a trademark in the UK?

Answer overleaf ➜

While you're thinking ...

'Aspirin' was (and remains in 80 countries) a
trademark of the German Bayer Corporation.

The name 'aspirin' is a contraction of the German
'acetylspirsäure', the name then given to acetylsalicylic
acid, which is the active ingredient in aspirin.

Acetylsalicylic acid was developed by Bayer in 1899, as an improvement
on the salicylic acid found in willow bark, which did act as a
painkiller but caused sharp pains and bleeding in the stomach.

Because Britain was granted the free use of the name 'aspirin' in the Treaty of Versailles

Originally only Bayer could make aspirin, as it had a trademark on the name, just as it did for its cough suppressant, heroin. However, aspirin proved something of a wonder drug as a painkiller and for reducing fever, particularly during the Spanish flu pandemic at the end of the First World War.

When the Treaty of Versailles was signed on 28 June 1919, it included a whole list of ways that Germany would be expected to pay in reparation for its acts during the war. Many of the contents of the list were the obvious ones: redrawing land boundaries, for instance, finally placing the much-disputed Alsace region firmly within France. There were restrictions on the build-up of arms, financial payments and requirements for industry. But also the UK and other countries gained the ability to use acetylsalicylic acid as 'aspirin' freely, without paying Bayer.

It might seem a very minor thing to include on such a game-changing treaty, but this simply demonstrates the significance of aspirin as a drug at this stage of its use.

Further reading: **Aspirin**

She sells sea salt on the seashore

How many grams of salt (sodium chloride) are there in a litre of typical seawater?

Answer overleaf ➜

While you're thinking ...

Salt, or sodium chloride, is probably one of the best-known chemical compounds – NaCl – and one of the very few inorganic substances we keep in our food cupboards.

We do need a small amount of salt in our diet, but we get more than enough from our food. Early humans probably got most of their salt needs from animal blood.

Despite the legend that seems to date back to Pliny, it is unlikely that Roman soldiers were paid in salt. Their salary (*salarium*, from the Latin *sal*, for salt) was more likely to be money with which to buy salt, or more generally for their food and other expenses.

There is no salt in seawater

Seawater contains no salt. Now, this seems ridiculous. It's even called salt water. Anyone who has had an accidental mouthful can tell you that seawater tastes intensely salty, and visit the right part of the world (the Isle de Ré in France, for example) and you can see mounds of white salt extracted from seawater. So what's going on?

In seawater there are all sorts of substances floating around, including both chloride ions (chlorine atoms with an extra electron) and sodium ions (sodium atoms with an electron missing). These two sets of ions come from totally different sources. The chlorine has mostly come from deep ocean vents, which pump out water that is rich in chlorine content. The sodium, meanwhile, will have been produced from rock based on sodium compounds like sodium silicate, dissolving in rivers and as a result of the pounding of the sea on the coastline.

Each of these separate substances is floating around in the seawater. It is only when the water is removed that the positive sodium and the negative chloride join together with electromagnetic attraction to form crystals of sodium chloride. The salt taste is from the sodium ions.

Once the salt is crystallised out, salt is salt – however it originated. So, despite the protestations of celebrity chefs (especially those who sell products) there is no difference whatever between the sodium chloride in sea salt and the sodium chloride from rock salt (or for that matter from reacting sodium and chlorine, though I wouldn't recommend this as it's a pretty violent reaction). Sea salt may have impurities, giving it a subtly different flavour, but all this does is make it less salty than the purer rock salt.

Further reading: **Inflight Science**

E is for blueberry

How many E-numbers would there be in a blueberry, if content labelling were extended to fruits?

Answer overleaf

While you're thinking ...

Blueberries are berries of a plant of the heather family, in the genus *Vaccinium*. The smaller fruit that is often called a European blueberry is actually the closely related bilberry.

Blueberries have tended to be labelled as 'superfruits' because they contain antioxidants, which were thought to have health benefits. However, it has since been shown that consuming antioxidants has no positive effect and can be bad for your health in large quantities. (If you want healthy fruit, cranberries and raspberries are significantly better, as they have half the sugar of blueberries.)

E-numbers are a system devised in Europe in 1962, which now provide codes for colourings, preservatives, antioxidants, emulsifiers and other food additives. The presence of a range of E-numbers is often taken as indicating that a food is unhealthy.

There would be at least 21 E-numbers in a blueberry

Score a point for anything between 20 and 30. Purists might argue that the answer should be zero, as E-numbers are additives and these chemicals are there already, but this is nit-picking.

If we had a contents list for a fruit like a blueberry, it would be scary indeed. They are 84 per cent water – which we'd think a bit of a rip-off in a manufactured product – and 10 per cent sugar, so mostly sugary water. They also contain nasty chemicals like the poison oxalic acid and some evil-sounding substances like styrene, benzaldehyde ... and ash. Not the kind of thing that you would want to feed to children. They contain both poisons and carcinogens.

What we forget in the comparison of natural and manu-factured foods is that practically everything we eat has toxins and other undesirable chemicals in it. Plants, for instance, tend to contain natural pesticides that are just as deadly as the artificial ones – but with poisons it is always the dose that counts, and the amount here is tiny.

It's also true that in the panic over E-numbers, we tend to forget that they cover a whole range of substances, even including, for instance, some vitamins, and plenty of nat-urally occurring chemicals. It might be helpful to put the traditional manufacturers' contents list into context if all foods carried them.

Further reading: **Science for Life**

FIRST SPECIAL ROUND: FAMOUS SCIENTISTS

Identify the scientist from his/her photograph or Nobel citation:

1.

2.

3.

4.

5.

Nobel citations

6. For the discovery of the Exclusion Principle.

7. For her determinations by X-ray techniques of the structures of important biochemical structures.

8. For his services to theoretical physics, and especially for his discovery of the photoelectric effect.

9. For his research into the nature of the chemical bond and its application to the elucidation of the structure of complex substances.

10. For their discoveries concerning the molecular structure of nucleic acids and their significance for information transfer in living material (three names).

Famous Scientists: Answers

1. Niels Bohr

2. Albert Einstein

3. Charles Darwin

4. Marie Curie

5. Ernest Rutherford

6. Wolfgang Pauli

7. Dorothy Hodgkin

8. Albert Einstein (again)

9. Linus Pauling

10. Francis Crick, James Watson, Maurice Wilkins

QUIZ 1
SECOND
SPECIAL ROUND:
CRYPTIC SCIENCE

Decode this cryptic message using scientific abbreviations and symbols:

1. $\sqrt{-1}$

2. Speed of light symbol

3. Uranium symbol

4. E/h in quantum physics

5. Helium symbol

6. Cube root of 512

7. Thorium symbol + Energy symbol

8. π

Bonus of two points for getting all of them correct and producing a meaningful sentence.

Cryptic Science: Answers

1. i

2. see (c)

3. you (U)

4. knew (from the formula for the energy of a photon E = hv, where h is Planck's constant and v is the frequency. v is the Greek letter nu, pronounced 'knew')

5. He

6. ate (8)

7. The

8. pie (pi)

Bonus of two points for getting the whole sentence.

QUIZ 2
ROUND 1: PHYSICS

Isaac's apples

Newton was inspired to muse on gravity by an apple falling on his head. True or false?

Answer overleaf ➜

While you're thinking ...

You can visit Newton's old home in Woolsthorpe, Lincolnshire, courtesy of the National Trust, and even see the famous apple tree.

The tree survived a near disaster when it partly collapsed during a storm in 1820, but more recently it has had to be protected by a willow barrier to stop tourists killing it off.

The apple tree is said to be about 400 years old, so would have been relatively new when Newton was born on Christmas Day 1642. (Now, confusingly, just in 1643, due to calendar reform.)

False – an apple never fell on Newton's head

It is a myth that an apple fell on Newton's head. But it is a myth inspired by a story straight from the horse's mouth. Newton told the story to antiquarian William Stukeley in 1726. Stukeley noted:

> After dinner, the weather being warm, we went into the garden, and drank tea under the shade of some apple trees; only he and myself. Amidst other discourse, he told me, he was just in the same situation, as when formerly, the notion of gravitation came into his mind. Why should that apple always descend perpendicularly to the ground, thought he to himself; occasion'd by the fall of an apple, as he sat in a contemplative mood.

So what did he contemplate? Luckily we are told that as well:

> Why should it not go sideways, or upwards? But constantly to the earths center? Assuredly the reason is, that the earth draws it. There must be a drawing power in matter. The sum of the drawing power in the matter of the earth must be in the earths center, not in any side of the earth. Therefore does this apple fall perpendicularly, or towards the center. If matter thus draws matter; it must be in proportion of its quantity. Therefore the apple draws the earth, as well as the earth draws the apple.

You can see Stukeley's original manuscript here: royalsociety.org/library/moments/newton-apple

Further reading: **Gravity**

Time flies

What is the best real time machine we have built so far?

Answer overleaf ➜

While you're thinking ...

There is nothing in the laws of physics that prevents time travel.

According to special relativity, whenever something moves, time runs slowly on it, seen from surroundings that aren't moving the same way.

A frequent flyer crossing the Atlantic once a week for 40 years would have moved around 1/1,000th of a second into the future.

Our best time machine is Voyager 1

As we've seen, according to special relativity, if something is moving with respect to Earth, then its clock will run slower when we compare it with ours, meaning that (if it ever came back) it would have moved into its future.

What we need, therefore, to make a good time machine is something that has travelled as quickly as possible, ideally for a long period of time. Voyager 1 was launched in 1977 to study Jupiter and Saturn, and was then programmed to head out of the solar system. At the time of writing, it is just about leaving the system (we can't say for sure, as there is no exact definition of the boundary). Because Voyager 1 travels at around 17,000 metres per second (38,000 miles per hour) and has been on its way for decades, it has become a real time machine. So far, it has moved about 1.1 seconds into the future.

We are used to time machines like Doctor Who's TARDIS that disappear from one point in time and appear at another. But Voyager has never disappeared 'into the future'. Instead, if we had super-powerful telescopes we could see that a clock on Voyager was showing a time 1.1 seconds behind Earth time. If there were people on Voyager, they would have experienced 1.1 fewer seconds than we have on Earth. This isn't just an effect. They would genuinely be younger.

If this is hard to imagine, let's say instead that a spaceship had moved fast enough that its clock was 100 years behind Earth time. When the occupants returned, 100 extra years would have elapsed on Earth – they would have moved 100 years into their future.

Further reading: **Build Your Own Time Machine**

QUESTION 3
Somewhere inside the rainbow

What colour would have been missing from Newton's rainbow if the colours had been described 100 years earlier?

Answer overleaf ➔

While you're thinking ...

Newton identified seven essential colours in the rainbow, though
he was aware that there were many more intermediate colours.

Not every colour is in the rainbow. Magenta, for instance, formed by
mixing red and blue light, is not to be found anywhere in the spectrum
of light and so technically does not exist as a colour of light.

Rainbows are seen when sunlight passes into a collection of
raindrops. The light is refracted on the way in, is reflected from
the back of the drop and refracts again on the way out.

Orange would have been missing from Newton's rainbow

Isaac Newton was the first to realise that white light is made up of all the rainbow spectrum and also was the first to explain why an object appears to be a particular colour in white light, because it absorbs some of the colours and reflects (we would now say re-emits) other colours, forming the colour we see.

When describing the spectrum of sunlight passed through a prism, Newton came up with the traditional seven colours of the rainbow: red, orange, yellow, green, blue, indigo and violet. Looking at an actual rainbow or spectrum, it is clear that this division into seven is arbitrary. It is difficult to see more than five or six colour blocks, and in detail there are far more colour gradations.

This was apparent to Newton, but he seems to have wanted seven main colours, probably to match up with the seven musical notes, A to G. Because light and sound were both detected by senses, and both had a kind of 'spectrum', it seemed likely to those of the period (if not to us) that they had similarities.

However, Newton was lucky because 100 years earlier, the colour 'orange' did not exist. It was the name of the fruit, but had not been assigned as a colour name in its own right. Things we would now consider orange, like a pumpkin, were then described as red, with variants like scarlet used for 'redder' reds. In the end, colour distinctions are language-dependent – in Russian, for instance, dark blue and light blue are two distinct and differently named colours.

Further reading: **Light Years**

QUESTION 4
The electronic professor

How many electrons did J.J. Thomson's original 'plum pudding' model predict that a hydrogen atom should have?

Answer overleaf ➜

While you're thinking ...

British physicist Sir Joseph John Thomson, usually referred to as J.J., discovered the electron, first found in cathode rays, and realised that this was a component of the atom.

Manchester-born Thomson attended Owens College, which later became Manchester University, and went on to be Cavendish Professor of Physics at Cambridge. He won the Nobel Prize in 1906.

Thomson insisted on calling electrons 'corpuscles', though the name 'electrons' had been mooted several years earlier.

Thomson predicted 1,837 electrons in the hydrogen atom

Allow yourself a point for anything between 1,800 and 1,870. The reasoning was simple, if fatally flawed. In the original version of this model, which conceived of the atom as a collection of negative electrical charges (the plums) scattered through a positive matrix (the pudding), that positive matrix had no mass. This meant that the entire mass of the atom had to be made up of electrons.

We now know that a proton's mass is around 1,836 times that of the electron. So that meant that Thomson would have needed 1,836 electrons to provide the mass of the single proton that makes up a hydrogen nucleus, plus one more for the atom's sole real electron.

Thomson's model did not last very long. Soon after, Ernest Rutherford's team at Manchester discovered the atomic nucleus, and realised that the majority of the mass was in this compact core, rather than in hundreds of imagined electrons.

*Further reading: **The Quantum Age***

QUESTION 5
Beam me up

If you could process a billion atoms per second, how long in years would it take to teleport a typical human being?

Answer overleaf ➜

While you're thinking ...

Quantum teleportation, a process based on quantum entanglement, makes it possible to transfer properties from one particle to another, which can be at a distance. This, in effect, turns the remote particle into the original one.

Unfortunately, quantum teleportation can't handle a whole person at one go, so it would have to work with relatively small collections of atoms at a time, hence the question.

Teleportation wouldn't be a very pleasant way of travelling. To achieve this transfer, the original particle has to, effectively, lose its identity. You would end up with a perfect copy, but the original 'you' would be disintegrated.

It would take 200 billion years to teleport a human being

Allow yourself a point for anything between 150 billion and 300 billion years. The trouble is that with any reasonable number of atoms transferred per second, teleportation will take too long. This is because there are so many atoms in a human body – typically around 7×10^{27}, though obviously it varies significantly depending on the size of the body. That's 7,000 trillion trillion – rather a lot to work your way through.

This doesn't mean that quantum teleportation is useless, though – in fact, it is essential for quantum computers. These are computers that use quantum particles like photons of light or electrons as the 'bits' in the machine. The problem with quantum particles is that if you make a measurement, you change the value – so you can't just examine a quantum particle to see what state it is in. But by using quantum teleportation, the state of a particle can be transferred from place to place and made to interact with other particles to perform a calculation.

It isn't absolutely impossible that quantum teleportation could be used to transfer very small objects from place to place, but for most of our needs, where we don't need to-the-atom accuracy, something like a 3D printer proves a much more practical way of 'sending' a physical object from A to B without ever moving it, by simply transferring the data that describes it.

*Further reading: **The Quantum Age***

Balletic bullet ballistics

If you fire a bullet horizontally and drop one from your hand at the same height and time, which will hit the ground first?

Answer overleaf

While you're thinking ...

Travelling at around 600 miles an hour, the 0.32-inch rimfire bullet from a Smith & Wesson number two revolver would cover 268 metres in one second.

The acceleration due to gravity at sea level is around 9.81 metres per second per second.

The curvature of the Earth has sufficiently little impact not to be of great concern here.

Neither bullet will hit the ground first

Both bullets will reach the ground at the same time. This seems counterintuitive, but both bullets have exactly the same force pulling them towards the Earth – gravity – and accordingly will accelerate towards the ground at the same rate.

There are two reasons why this doesn't seem likely. One is that we know it will take a dropped bullet only around a second to hit the ground, but we expect a bullet from a gun to go further than that time allows. The other is that we suffer from cartoon-character physics. In a cartoon, a character who runs off a cliff keeps going in a straight line until he or she notices what has happened and begins to fall. We have an equivalent feeling that the bullet, shooting forward at high speed, should somehow be able to stay up in the air longer.

Of course, some objects can keep airborne for longer if they are moving forwards and if, for instance, they have wings that provide lift as a result of the air moving over them and being deflected. But bullets are not known for having wings.

The best way to predict what will happen to something moving is to look at the forces acting on it. When, for instance, you throw a ball up in the air, ignoring air resistance, the only force acting on it is gravity, acting downwards. This is true from the moment the ball leaves your hand, through the point where it stops at the top of its trajectory, on to its plummet to Earth. There is just one force acting in one direction.

Further reading: **Gravity**

Moonshine power

Which physicist said: 'The energy produced by the breaking down of the atom is a very poor kind of thing. Anyone who expects a source of power from the transformation of these atoms is talking moonshine'?

Answer overleaf

While you're thinking ...

This famous quote on the limited nature of nuclear energy was printed in a US newspaper in the early 1930s.

The first nuclear transmutation, changing nitrogen into oxygen through fission, had been undertaken in 1917.

This was taken a step further in 1932, when lithium-7 underwent fission in a bombardment of accelerated protons. It was then that the term 'splitting the atom' was popularised.

Ernest Rutherford said that energy from fission was moonshine

The irony here, of course, is that not only was Ernest Rutherford a great physicist, he and his team were responsible for the discovery of the atomic nucleus in his Manchester laboratory in experiments between 1908 and 1913. Rutherford is sometimes referred to as the 'father of nuclear physics'.

Rutherford was quoted in the *New York Herald Tribune* of 12 September 1933, timing that was particularly striking because this was the same year that Leo Szilard proposed the idea of a nuclear chain reaction. Without a chain reaction, Rutherford was, effectively, correct – but Szilard's idea would transform production of energy from nuclear fission. In less than ten years from Rutherford's pronouncement, the first nuclear reactor, located under an old stadium in Chicago, went critical.

History has shown time and again that great scientists are no better at predicting the future than the rest of us, because the most important new developments tend to come from unexpected directions. But Rutherford, who certainly had more insight than most, was particularly unlucky with his timing.

Further reading: **Nuclear Power**

Gamow's game

Who did George Gamow lie about being a co-author with him and Ralph Alpher? And why?

Answer overleaf →

While you're thinking ...

Russian-born physicist George Gamow was a respected theoretical physicist and cosmologist who worked on a number of topics, including stellar nucleosynthesis (the formation of atoms in stars) and the Big Bang.

Gamow was also a respected science communicator who wrote a number of books for the general public, including the charming 'Mr Tompkins' series, which examined what the world would be like if, for instance, light travelled much slower.

Ralph Alpher was a significantly younger cosmologist.

Hans Bethe. So the paper's authors would be Alpher, Bethe and Gamow

Both answers are required for the point. Known for his sense of humour, as soon as George Gamow realised he would be writing a joint paper with Ralph Alpher he wanted to rope in Bethe, to get their names spelling out something close to alpha, beta, gamma, the first three letters of the Greek alphabet.

Hans Bethe, who was a respected elder statesman of physics, played no part in writing the paper. Alpher was not happy with this, as he was significantly junior to the other two and felt that he would be overlooked as an author, but he was persuaded to go along with it, and the paper is generally known as the Alpha-Bethe-Gamow or the αβγ paper.

The paper is a significant one, outlining the way that the Big Bang could have produced the lighter elements, although it was incorrect in assuming that all early elements were produced this way.

Alpher continued to express his dislike for the joke for the rest of his career. Given that the joke is widely known, it seems unlikely that he was assumed to be a junior associate to the other two, but it has perhaps overshadowed his later work.

Further reading: **A Brief History of Infinity**

QUIZ 2
ROUND 2: BIOLOGY

QUESTION 1
Tasty teaser

Which part of the tongue contains the salt-detecting taste buds?

Answer overleaf

While you're thinking ...

Our taste buds handle (at least) five types of taste:
salt, sweet, sour, bitter and savoury.

The human tongue has in the region of 3,000 to 8,000 taste buds.

Taste buds are typically replaced every couple of weeks.

All parts of the tongue can taste salt

If you've suggested that there are particular regions of the tongue, for instance patches at either side towards the front, that deal with 'salt', then you are perpetuating a myth – take a one-point penalty.

The idea that taste buds are grouped on specific areas of the tongue dates back to the start of the 20th century when a map of the tongue showing such zones that were more or less sensitive was misinterpreted. It's hard to tell just how this myth got so widespread, but I remember doing an experiment in junior school that attempted to demonstrate this – different areas of the tongue were supposed to produce different taste sensations when prodded.

If, as we did, you tried this out and confirmed the theory, you were instead demonstrating a common error in scientific investigation where the investigator expects a particular outcome, and this biases his or her results to produce the desired effect.

Whatever the reason, the idea became popular, but the map of the taste areas of the tongue is totally bogus.

*Further reading: **The Universe Inside You***

Our mousy cousins

How long ago did our most recent common ancestor with mice live?

Answer overleaf

While you're thinking ...

All terrestrial organisms are likely at some point in history to have had a common ancestor.

Mice and humans share about 70 per cent of the same protein-coding gene sequences.

Mice are often used in laboratory experiments, in part because they are cheap and have short lifespans, and in part because they are mammals with a relatively high share of genes with humans (known as homology).

Our common ancestor with mice lived around 75 million years ago

You can have a point for anything between 70 and 80 million years.

There's a paradox in evolution that every organism is the same species as its parents, yet go far enough back through an ancestry and you will find a totally different species. It's a bit like splitting a colour spectrum down to millions of segments. Each segment will be, to all intents and purposes, the same colour as adjacent ones, but you still get the whole spectrum from red to violet when you stand back and take a look at the whole.

It should be obvious, but when comparing ourselves with another species, we should not equate the common ancestor with either us or the other species. So, for instance, if you look at our common ancestor with chimpanzees, which flourished somewhere between 7 and 20 million years ago, it's not a human and it's not a chimpanzee. Both species have evolved since – though the change in DNA is surprisingly small. We differ from chimpanzees only in about 0.3 per cent of that coding DNA.

Similarly, although the common ancestor with mice, living back when dinosaurs were still at large, would look more like a mouse than a human, it was no more a mouse than we are.

While in principle there could have been several examples of life that started separately, each producing its own evolutionary family, the common aspects of DNA and RNA that we find in all living things suggests that we probably all started from a single common ancestor.

Further reading: **The Universe Inside You**

Can a ground lion change its spots?

Why does a chameleon change colour?

Answer overleaf ➜

While you're thinking ...

Chameleons are lizards with long tongues that can be
shot out at high speed to capture prey, and individually
mobile eyes, often in a turret-like structure.

Chameleons aren't just found in forests, but also in desert
environments – warmth is their main criterion.

For no really obvious reason, the name 'chameleon' means ground lion.

Chameleons change colour for heat management and signalling

Score a point if you said either heat management or signalling. Have a bonus half-point if you said both. But take away a point if you said concealment.

It's ironic that the chameleon, which is often used as the definitive example of concealment by changing to resemble its background, doesn't actually do it. Some species of chameleon can, indeed, perform colour changes. They have special cells in their skin called chromatophores, which act a little like the pixels on an LCD screen. But they don't use the colour variation to hide.

For the real 'chameleon effect' you need not a chameleon but one of a number of flatfish, which really can perform this trick. The advantage they have over the chameleon is that, as well as having chromatophores on their upper surface, they also have light-sensitive patches on their lower side, which can transmit an approximation of the look of the seabed's colouring to the fish's upper surface.

Probably the simplest examples of human-made cloaking devices work the same way – by having a bank of cameras behind the object to be concealed, and then projecting the image from the front of the object. But it is much harder to get the effect to work convincingly in this situation. When a flatfish lies on the seabed, there is practically no perspective shift between the top of the fish and the sand. But because there is usually a considerable gap between the object being concealed by a human-made cloak and its background, there are often parallax problems, where the viewed object moves oddly with respect to its background.

Further reading: **Ten Billion Tomorrows**

Making sense

How many senses do humans have?

Answer overleaf →

While you're thinking ...

There are plenty of senses we don't have that other organisms do. For example, some birds appear to be able to detect the Earth's magnetic field, while some aquatic creatures, like sharks, can detect the electrical fields emitted by the nervous systems of their prey.

Other organisms make use of the familiar senses in different ways. Some, for instance, can see infrared or ultraviolet, while bats use their extraordinary auditory echo-location system to navigate.

Sensory input is usually modified by the brain to produce an unreal but helpful model. So, for instance, what we see is a construct from the input of optical nerves, not a camera-like picture. Other animals may have different senses integrated into such a picture – dogs, for instance, may have their sense of smell more strongly integrated with sight than we do.

We have a lot more than five senses

There isn't a precise number that's recognised, but give yourself a point for anything more than five, because it certainly isn't just that many. Think, for example, of the sense that tells you which way up you are. Or proprioception, the sense that gives you an awareness of the location of parts of your body, so you can touch your nose without looking.

For that matter, we have an infrared sense – not as good as organisms that have it built into their sight, but our skin has crude infrared detectors so that you can tell, for instance, if an iron is switched on when it is several centimetres from your hand. And then, like a smartphone, you have an acceleration sensor, in this case in your inner ear, that makes an important contribution to your sense of balance. So that's at least nine senses.

One extra sense that many would add is the sense of pain. The burning sensation when biting into a strong chilli, or the pain you feel when a needle is pushed into your arm, seem related to taste and touch respectively – and certainly they are transmitted by the same receptors. But arguably there is a significant difference between the warning sense of pain and the other uses those sensors are put to. You can also, of course, experience internal pain – and most of us tend not to think of touch or taste being a sense that is attributed to our internal organs.

Further reading: **The Universe Inside You**

Daily decay

How long is the half-life of DNA, and what are the implications for Jurassic Park?

Answer overleaf →

While you're thinking ...

A half-life is a measure of the rate of decay of a substance. Most commonly applied to radioactive material, it is sometimes also used for the breakdown of organic material.

DNA, or deoxyribonucleic acid, provides the 'construction information' for biological cells.

Jurassic Park, first a 1990 novel by Michael Crichton, then filmed to huge commercial acclaim by Steven Spielberg in 1993, is based on the hypothesis that dinosaurs could be recreated from fossil DNA.

The half-life of DNA is 521 years – *Jurassic Park* is a non-starter

Have one point for anything between 500 and 550 years. And a bonus half-point if you said that the implication was that *Jurassic Park* couldn't happen for real.

The premise of *Jurassic Park* was indubitably very clever. Blood-sucking insects have been found trapped in amber, which is fossilised tree resin, dating back millions of years. If one of those insects happened to have taken blood from a dinosaur, then perhaps dinosaur DNA could be extracted, and by using cloning techniques, dinosaurs could be brought back to life.

However, when researchers tried to extract DNA from blood preserved this way, they failed. It turns out that DNA has a half-life – and this has been shown to be around 521 years. So in a well-preserved sample, around half the DNA would still be present after that time period. That gives us limits on the survivability of any DNA. It suggests that DNA would be totally unreadable after around 1.5 million years and totally gone within a maximum of 6.8 million years, far short of the 65 million-year gap to the time of the dinosaurs.

There is, however, still hope for mammoths. Mammoths have been found well preserved in ice, and have far closer relatives still living to act as a host, compared with dinosaurs. And since mammoths lived up to around 5,000 years ago, there is every possibility of finding some usable DNA. At the time of writing, two projects to clone a mammoth are under way, though neither has as yet achieved results.

Further reading: **Ten Billion Tomorrows**

QUESTION 6
Taste-tingling tangs

What is the formal name of the savoury taste in the five main taste groups?

Answer overleaf ➔

While you're thinking ...

Tastes are the result of detectors on the surface of the tongue reacting with different chemical constituents in a substance. 'Salt' taste, for instance, is a reaction to the presence of sodium or potassium ions.

The five basic tastes are no more unique than the seven colours of the rainbow – we experience a continuous response to the different chemical stimuli on the tongue, though these five elements do contribute to this.

Some culinary cultures suggest that there is a sixth taste group of 'pungency' or 'hotness', as in the response to foods like chilli, but in fact this is a rather different kind of response to the other tastes.

The savoury taste is called umami

It might seem odd to make use of this Japanese word (meaning something like a pleasing savoury taste) in what is otherwise a list of familiar English terms, but umami is used in part because it was proposed by a Japanese chemist and in part because it is triggered by glutamates (and certain other organic compounds) which are often added as taste-enhancers to Asian cuisine.

Although we are probably most familiar with the name from monosodium glutamate (MSG), glutamates are common in a range of foods, which is one of the reasons that the so-called 'Chinese restaurant syndrome' – chest pains, headaches, sweating, etc. after eating a Chinese meal, and associated with consuming MSG – is unlikely to be anything more than a nocebo, which is a negative placebo. Glutamates occur widely in tomatoes, yeast products and Parmesan, and are also produced inside the body, which makes them unlikely culprits.

Other umami stimulators are found in fish, cured meats and some vegetables.

Further reading: **The Universe Inside You** *and* **Science for Life**

Dolly mixture

Why was Dolly the sheep called 'Dolly'?

Answer overleaf ➜

While you're thinking ...

Dolly the sheep, born in 1996, was the first mammal to be cloned.

Out of 276 eggs in Dolly's trial, hers was the only one to survive.

Since Dolly, other mammals to be cloned include
cats, dogs, pigs, cows, wolves and camels.

Dolly was named after Dolly Parton, as she was cloned from a mammary cell

Give yourself half a point for Dolly Parton and half for the mammary cell. Dolly's parent ewe was long dead – the animal was cloned from a cell that had been kept alive in the laboratory as a culture.

Dolly appeared to be a perfectly normal sheep. She did die relatively young, but despite rumours, this seems to have been the result of a common sheep disease, rather than a connection to her being a clone.

However, the success of Dolly, and a whole range of subsequent mammals, does not mean that we will see human clones any time soon. (There are a small number of claims to have done this, but none has been verified.) All too often, cloning fails after the embryo has started to form, or when the animal has been born, which would be unacceptable for human cloning.

The process does also seem significantly more complicated with apes than with sheep or cats or dogs. It's easy to damage genes in the cloning process, which is a little like trying to repair a watch with a hammer and chisel. Until the process is made a lot more reliable, it would not be ethical to attempt a human clone.

Further reading: **Ten Billion Tomorrows**

QUESTION 8
Mouse menace

How does a hawk spot mice in the grass while hovering high above?

Answer overleaf

While you're thinking ...

The small hawk (or more properly, the falcon) most frequently
seen floating on the rising air by the banks of roads in
the UK, looking for mice and voles, is a kestrel.

There are at least twenty species of field mouse, each
distinct from the common house mouse.

Pet mice and lab mice are usually just house mice, *Mus musculus*.
There is a wide range of special type variants of the lab mouse,
each with their own code name (e.g. BALB/c) and specific genetic
make-up. This code is set by a body known as the Committee
on Standardized Nomenclature for Inbred Strain of Mice.

The hawk follows a trail of pee

Equally acceptable as an answer is 'by seeing ultraviolet'. Humans have four types of light sensors in their eyes – rods, which handle black and white, and three types of cones, one type most sensitive to each of the three primary colours of light: red, blue and green. This enables us to see the usual visual spectrum from red to violet – but in reality, the full spectrum of light stretches far below, to take in radio waves and microwaves, and far above, up into X-rays and gamma rays.

Either side of the visual spectrum are infrared just below visible light and ultraviolet just above. But when we say 'visible light', we really mean 'visible-to-human' light. In the case of that hawk, hovering over a grassy bank, a fourth type of cone is available that is sensitive to ultraviolet.

So, imagine the scene. The hawk is soaring high above, near stationary in the sky, and below a mouse is scampering along, pretty well disguised by the brown-green undergrowth of the grassland. The chances are, the hawk will never spot it. But its small prey – mice, voles and shrews – fairly constantly emit small quantities of pee, leaving a trail behind them. And that urine glows in the ultraviolet light from the Sun. So the hawk gets a glowing 'here I am' flag from its prey.

Further reading: **The Universe Inside You**

QUIZ 2
ROUND 3: THE MIX

QUESTION 1
Train drain

What did Emmett Leith and Juris Upatnieks do in 1964 using a model train and two stuffed pigeons?

Answer overleaf ➔

While you're thinking ...

The train at the heart of this question was an American steam engine.

It was sometimes accompanied by a larger-scale model train signal and a smaller-scale toy of a person riding a bike.

The engine was called *The General.*

They used them as the subject of the first hologram

Working at the University of Michigan, Leith and Upatnieks produced the first viewable holographic images of a motley collection of objects.

The concept was dreamed up by Hungarian-British scientist Dennis Gabor in 1948 when he was looking for a better way to capture the information from an electron microscope, but the technology did not then exist to make it possible. What was required was a source of coherent light – monochrome and with the photons in phase. This was developed in 1960 in the form of the laser, and just four years later the first hologram was produced.

Where a conventional photograph simply captures the intensity at various points on the image, a hologram like Leith and Upatnieks' uses the interference between a pair of lasers, one used to illuminate the subject and the other hitting the recording medium directly. The result is an interference pattern, which holds information about both intensity and the phase and direction of the incoming light. When the plate is illuminated with a laser, the result is an image that changes when looked at from different directions, just like a real view.

The other interesting aspect of the hologram is that every piece of it contains information on the whole view. Break off a piece and you can still see the whole picture, although with reduced clarity. When the first hologram was produced, it was thought they would rapidly become commonplace, but as yet the technical requirements for production and viewing mean that they remain novelties or used in specialist applications.

Further reading: Ten Billion Tomorrows

Light fantastic

What is the speed of light in metres per second?

Answer overleaf ➔

While you're thinking …

Many early natural philosophers thought light travelled instantly.

Galileo tried to measure light speed by timing a lantern flashed
at him from a few miles away. He failed miserably.

The first successful attempt to measure light speed
used observations of the moons of Jupiter.

The speed of light is 299,792,458 metres per second

The value for the speed of light is often given as 300,000,000 metres per second or 186,000 miles per hour, which is worth half a point, but the value above is needed for the full point. Usually if a quiz book publishes the exact value of a measurement like this it is dangerous, because over time it will be measured more accurately. But not light speed.

The speed of light was first measured in 1676 by Ole Rømer, a Danish astronomer working in Paris, who took regular measurements over many months of the positions of the moons of Jupiter. The moons spent a long time gradually speeding up, then, for another long period, slowing down. Rømer realised that the shift in timings reflected the different amount of distance the light from the moons had to cover as Jupiter got nearer to and farther from Earth. He made the speed of light around 220,000,000 metres per second – not bad considering the technology available. Over the years, the speed was measured more and more accurately. Measurements moved from space into the laboratory, first using toothed wheels and rotating mirrors, then electronics, until 1983. In that year it became impossible to measure the speed of light more accurately.

That sounds weird. By 1983 we had a very precise definition of a second, but the definition of a metre wasn't good enough. It was decided to *define* the metre as 1/299,792,458th of the distance light travels in a second. So the speed of light in a vacuum will remain 299,792,458 metres per second. Exactly what a metre is will vary a touch as our measurement of a second is made more accurate – but the light speed won't change.

*Further reading: **Light Years***

QUESTION 3
Alphanumeric enigma

What is the significance of the sequence of characters 6EQUJ5?

Answer overleaf ➜

While you're thinking ...

Letters are quite often used to extend the numerical range,
for example in the hexadecimal arithmetic employed by some
computers, the character set 0123456789ABCDEF is used.

In such a hexadecimal representation 42 would
be 2A, while 2015 would be 7DF.

The sequence 6EQUJ5, which features in this question, was printed as
part of many other characters on a long stack of computer printout.

The sequence 6EQUJ5 is the 'Wow signal' detected by SETI

Have a point for 'Wow signal' and a bonus half-point (which you can have without the 'Wow signal' for mentioning radio telescope evidence for extraterrestrial life).

In 1977 a volunteer, Jerry Ehman, was using the Ohio State University 'Big Ear' radio telescope to monitor radio signals from space as part of SETI – the Search for ExtraTerrestrial Intelligence.

Read-outs from radio telescopes in the early days often consisted of long printouts, with strings of characters, each denoting the strength of the signal received. The code started at a blank for no signal at all and went up through the numbers and then some of the letters, peaking at U, which would be the equivalent of strength 30.

This was, frankly, rather boring work. But on 15 August 1977, Ehman spotted a sequence of digits that identified an intense pulse – 6EQUJ5. The pulse lasted for around 72 seconds, which was the kind of period you would expect from a distant source as the fixed telescope scanned the skies with the rotation of the Earth. Ehman put a ring around the characters and wrote 'Wow!' in the margin, hence the name.

Despite intense efforts, no further activity from that direction, in the constellation Sagittarius, was spotted. As with all radio telescope anomalies, the first suspicion had to be that this was an Earth-based source, but the possibility that it was a signal from the stars – natural or artificial – has never been disproved.

*Further reading: **Ten Billion Tomorrows***

Bovril bonanza

What is 'vril' (as in Bovril)?

Answer overleaf ➔

While you're thinking ...

Bovril is a food product that comes in a jar. It appears
not unlike Marmite or Vegemite, but rather than being
based on yeast extract it has a meat source.

Bovril was first produced in the 1870s and is still
available in the UK. It was originally called Johnston's
Fluid Beef. 'Bovril' is probably an improvement.

Although Bovril is sometimes used as a spread like Marmite, or to add
flavour to gravy, it is uniquely also used with hot water to make a
drink ('beef tea'). This used to be popular from vending machines in
swimming baths when I was young, possibly because it was the only
hot drink these machines produced that had a detectable flavour.

'Vril' is a mysterious energy source in a novel by Edward Bulwer-Lytton

There's a point for 'mysterious fictional energy source' and a bonus half-point if you got the author's name as well.

The Victorian writer Edward Bulwer-Lytton is probably best known now as the author of what is generally considered the worst opening line in all literature: 'It was a dark and stormy night.' Later borrowed by Snoopy in the *Peanuts* cartoon strip, this immortal line has spawned the Bulwer-Lytton Fiction Contest for the worst opening line of the year.

In his day, though, Bulwer-Lytton was a popular novelist, and in his science fiction epic *The Coming Race*, the protagonists made use of a form of 'atmospheric magnetism' that could be directed through staffs of power to become a deadly weapon. This mysterious source of energy was called 'vril'.

There was, of course, no science behind vril, but the concept became popular, cropping up in various crank religions, and a number of products incorporated the made-up word into their brand name in the hope that it would make them sound energy-giving and health-enhancing. (Similarly, in the early 20th century, radium would be used to make even as bland a product as toothpaste sound more effective – though with rather deadlier consequences.)

As far as I am aware, the only product still for sale with a 'vril' name is Bovril, which was a combination of vril with a contraction of 'bos', the Latin for a cow.

Further reading: **Ten Billion Tomorrows**

The Wardenclyffe wonder

What was Wardenclyffe?

Answer overleaf ➜

While you're thinking ...

Wardenclyffe was located in Shoreham, New York.

Wardenclyffe was briefly in the world news for science-connected reasons in the first years of the 20th century.

The name 'Wardenclyffe' is derived from that of property developer James S. Warden, who owned a large swathe of land in the area, intended for a resort development.

Wardenclyffe was the site of Nikola Tesla's experiment in wireless transmitted energy

Have half a point for Tesla and half a point for wireless transmission of energy.

There are few more controversial figures in the history of science than Serbian-born American inventor and engineer, Nikola Tesla. His fans think that he was a genius who eclipsed Edison as an inventor and Einstein as a scientist. In reality, he was a brilliant engineer, devising some of the key technology for the AC electrical system, as well as early radio-controlled devices and novel generators. But he was at best a poor scientist who, though he was brilliant at *using* electricity, did not understand modern electromagnetism theory and never accepted relativity.

Tesla believed that it was possible to transmit electrical power as a standing wave through the Earth, with a return circuit through the air. The Wardenclyffe development (portrayed impressively in the movie *The Prestige*) was a purpose-built structure with a 57-metre-high tower that was connected to high-voltage generators with the aim of being able both to communicate wirelessly to anywhere in the world and to broadcast electrical power.

There was indeed a localised electrical induction effect that meant anything metal nearby could have a current flowing in it (including the neighbours' taps) – but Tesla's imagined long-range waves didn't materialise, and despite financier J.P. Morgan pouring large amounts of money into the project, it was abandoned in 1903. The tower was scrapped, but the building remains.

Further reading: **Ten Billion Tomorrows**

Unpronounceable

Why did Uuq become Fl?

Answer overleaf →

While you're thinking ...

The *Oxford English Dictionary* contains no words between utu and uva.

'Fl' can be used as a shortening of the Latin 'floruit' meaning 'he, she or it flourished' to indicate the active working period of a person in history.

Unlike 'Uu' words, 'Aa' words are surprisingly common, including aa itself, which is both an obsolete term for a stream and a geological term for a rough surface covered in loose clinker. You can also have an aal, aam and aapa, among others, before arriving at the familiar aardvark.

HOW MANY MOONS DOES THE EARTH HAVE?   **167**

Because flerovium (Fl) is the element formerly known as ununquadium (Uuq)

Yes, it's not only Prince who can be 'formerly known as ...'

Element 114 in the periodic table was first spotted at Dubna in Russia. When new elements are detected, they tend to be given a holding name that reflects their position – ununquadium is just 'one-one-four-ium' in shortened Latin. And it was as Uuq that it was known for some time, and can still be seen on many printed periodic tables. The final name, flerovium, awarded by the International Union of Pure and Applied Chemistry in 2012, comes from the name of the laboratory in Dubna, the Flerov Laboratory, which itself is named after the nuclear physicist Georgi Flerov (or Flyorov).

In 1998, the team at Dubna fired high-energy calcium 48 ions at a plutonium 244 target. This went on for a total of 40 days, during which 5,000,000,000,000,000,000 ions were shot down the accelerator to hit the plutonium. Just one single atom of the isotope 289 of flerovium was discovered, which took 30.4 seconds to decay.

This relatively long time is because flerovium sits in an island of stability, a position in the periodic table where the atomic nucleus has a near-spherical configuration, which suggests that its half-life should be longer than that of its neighbours. Where, for instance, darmstadtium, which comes before this island of stability in the table, has a half-life typically measured in microseconds, flerovium 289, as we have seen, stays around for seconds at a time.

Further reading: ***www.rsc.org/periodic-table/podcast***

Hazardous hydration

Why are we recommended to drink eight glasses of water a day?

Answer overleaf ➔

While you're thinking ...

A 'glass' is not a well-defined unit, but usually refers to something between the large wine glass, at 250ml, and the typical soft-drink glass, at 330ml.

Water should not be consumed too quickly. Drinking several litres in one go, as some sportspeople have done, can result in swelling of cells, brain damage and even death.

When undertaking sports, despite claims from sports drinks manufacturers, there is no need to 'stay ahead of your thirst' – thirst has been shown to be an excellent guide to when you need to drink, so wait until you are thirsty.

Nobody knows why eight glasses of water are recommended

You can also have a point if you said 'We aren't' or similar. One of the best-known myths in dietary science, the idea that you need to drink eight glasses of water a day has become deeply embedded in our culture, but it has no basis in truth.

Although no one is quite sure where this myth that requires us to drink 2 to 2.5 litres a day came from, it may be derived from a 1945 US National Research Council recommendation that adults should consume a millilitre of water for each food calorie, which matches up with those 2 to 2.5 litre figures.

The reason that the recommendation is a myth is two-fold. Firstly the volume is misleading – because we will typically get half this amount of water from our food (which counts just as much as water in a glass), so we typically need to drink only around 1 litre a day. Secondly, the assumption is often that the fluid we consume needs to be literally water to meet this requirement, but in fact there is no difference in hydration benefits between water and, say, a soft drink.

Even tea and coffee are fine if you aren't drinking so much as to overdose on caffeine – although they and cola are mild diuretics, this doesn't seem to influence their ability to hydrate. The same goes for sports drinks – they too will hydrate as well as water. But no better.

So rather than 'eight glasses of water a day', it really should be 'four glasses of water-based liquids'. A significant difference.

*Further reading: **Science for Life***

Hothouse habitat

If there were no greenhouse gases in our atmosphere, what would be the average surface temperature of the Earth?

Answer overleaf ➔

While you're thinking ...

We're used to carbon dioxide being the bad guy of the environmental world, but carbon dioxide in the atmosphere is essential for plants and for the positive aspect of the greenhouse effect. It's only too much carbon dioxide that is a problem.

Other gases contribute significantly more per molecule to the greenhouse effect than carbon dioxide. Methane, for example, has 23 times the impact, but there is far less of it in the atmosphere.

Venus demonstrates the greenhouse effect gone horribly wrong. With a 97 per cent carbon dioxide atmosphere it has a blistering surface temperature.

The Earth's average surface temperature would be −18°C

Give yourself a point for anything between −15°C and −21°C. The greenhouse effect on the Earth, with the combined effect of the different greenhouse gases present, raises average temperature about 33 degrees. If there were no greenhouse effect, life as we know it, beyond bacteria, would not exist on Earth. So, of itself it's a good thing – it's just when it goes too far that it's a problem.

Oddly, the greenhouse effect has nothing to do with greenhouses – they don't actually work the same way. The greenhouse effect involves gases in the atmosphere acting a bit like a one-way mirror. Sunlight mostly shoots straight through on the way in. It is then re-emitted by the Earth, mostly as lower-energy infrared light. Some of this gets absorbed by the greenhouse gas molecules and then is re-emitted, scattering some of the infrared back towards the Earth and so warming it up further than it would be otherwise.

Greenhouse gases don't just give us a habitable climate. Carbon dioxide, specifically, is an essential for plants to grow, as it is absorbed from the atmosphere and the carbon is extracted to go into the plants' structure. Not only do animals benefit from the plants producing the waste product of oxygen that the animals need to breathe, but also the whole food chain is dependent on vegetation at the bottom. If you are a herbivorous animal, you need those plants – and if you are a carnivore, you need the herbivores that eat the plants.

We might not need any extra CO_2, but we'd certainly be in a mess without it.

Further reading: **The Universe Inside You** *and* **Ecologic**

QUIZ 2
ROUND 4: HISTORY

Birth of Bacon

In what century was Roger Bacon, famed for his emphasis on the use of experiments in science, born?

Answer overleaf ➜

While you're thinking ...

The English word 'bacon' for cured pork comes from the Old French 'bacon' and has been recorded as far back as 1330.

More recent famous Bacons include the British artist Francis Bacon and the American actor Kevin Bacon.

For no entirely clear reason, Kevin Bacon is used as an example of the 'six degrees of separation' hypothesis, where practically anyone who has worked in Hollywood can be linked to Bacon in a maximum of six steps.

Roger Bacon was born in the 13th century

Roger Bacon is often confused with the 16th-century philosopher Francis Bacon, who was one of the first to explicitly establish processes like induction as key to the scientific method. Apart from sharing a surname, the much earlier Roger Bacon also emphasised the importance of experiment in science, though his idea of experiment was much looser than the later form.

Roger Bacon was born in 1214 or 1220 and became an early member of the Franciscan order. He had a remarkable rollercoaster of a life – he was determined that it was essential to collect and publish ideas on science, but struggled as a member of an order that was going through an anti-intellectual period, during which they were banned from writing books.

Bacon got assistance from a French cardinal, who soon after became Pope. It seemed this was the opening Bacon needed and he wrote a proposal for the Pope for an encyclopedia of science. Bacon got so carried away writing this that it ended up over 500,000 words long. He made two other attempts at writing a covering letter that each became books in their own right before sending off the package to the Pope. Sadly, his sponsor died while the proposal was on its way to Italy and Bacon fell out of favour.

His books remain a fascinating portrayal of scientific knowledge of the period, including a considerable amount of original work from Bacon, especially on light, and ending with a whole section on the importance of experiment and maths to science. Bacon's story was later embellished with a host of legends before he became confused with his (as far as we know) unrelated successor.

Further reading: **Roger Bacon**

Albert's alma mater

Which university was Einstein working at when he wrote papers establishing the reality of atoms, laying the foundations of quantum theory, and describing special relativity?

Answer overleaf ➔

While you're thinking ...

Albert Einstein was born in Ulm in 1879, but would later renounce his German citizenship.

The head of physics at the Eidgenössische Technische Hochschule or ETH (Federal Institute of Technology), in Zürich once told Einstein: 'You're a very clever boy, but you have one big fault. You will never allow yourself to be told anything.'

When Einstein wrote his 1905 paper on Brownian motion, many scientists believed that atoms didn't exist, but rather were a useful metaphorical concept.

Einstein was not at university when he wrote his celebrated papers

In 1905, when Einstein published his three remarkable papers – including the paper on the photoelectric effect that was one of the starting points of quantum theory (and won him the Nobel Prize), and the paper that formulated special relativity – he was an amateur without a university post. He was working in the Swiss Patent Office in Bern as a clerk.

It has been suggested that his work there had a direct impact on his development of special relativity. At the time, there was a lot of interest in coordinating electric clocks at different locations. Until the coming of the railways, time was set loosely city by city – you might find midday in, say, Bern, came half an hour later than midday in Basel. But when a railway linked them, it wasn't acceptable for the times to vary, and so 'railway time' made it necessary to synchronise clocks in different locations.

The whole idea of when two events are simultaneous is central to the thinking behind special relativity – and interestingly Einstein makes use of a moving train as a way of illustrating how light's behaviour makes simultaneity a relativistic concept. So it doesn't seem too far-fetched that Einstein was indeed inspired in his thinking by some of those patents.

Einstein finally got an academic position at Zurich University in 1909 – the year in which he gave his first-ever paper at an academic conference, which included that most famous equation $E = mc^2$. By this time he was 30, an age by which many scientists have already done their best work – but it was only the beginning for Einstein.

Further reading: **Einstein**

Cloud conundrum

What type of cloud is 'cloud nine' – and why is this used in a saying?

Answer overleaf ➔

While you're thinking ...

The saying in question, 'I'm on cloud nine', indicates
that you are happy and content.

There are at least 52 different recognised types of cloud.

The original classification of clouds, dating back to 1802,
had three families: cirrus, cumulus and stratus.

Cloud nine is a cumulonimbus – used in the saying because it's the highest cloud type

A half-point for each part. Some forms of cumulonimbus are known as thunderheads – the huge, anvil-shaped clouds that are responsible for many thunderstorms. They are very deep clouds; though starting off low to the ground at the base, they reach as far as 18 kilometres into the air. In the cloud numbering system developed at the end of the 19th century, which ran from one to nine, cumulonimbus were assigned number nine, because they reached such a significant altitude, towering above the other clouds.

Because of this, even though a cumulonimbus can be extremely unpleasant to experience as a source of a storm, 'being on cloud nine' came to mean being on top of the world – perhaps because of the traditional association of heaven with the heights of the skies. If you were on cloud nine, you were indeed happy and content.

This was fine, until the World Meteorological Organization (WMO) decided that having nine basic types wasn't enough, and renumbered the clouds, making the highest cloud number ten. This was perfectly sensible from a scientific viewpoint, but it spoiled the saying.

In an unusual gesture, the WMO realised that they were being spoilsports and soon after renumbered the clouds to run from zero to nine, still providing ten types, but ensuring that once again the highest cloud, on top of the world, was cloud nine.

*Further reading: **Inflight Science***

QUESTION 4
Passing the baton

Newton was born the same year that Galileo died. True or false?

Answer overleaf

While you're thinking ...

Galileo died in January 1642.

Newton's father died before young Isaac was born at Woolsthorpe in Lincolnshire.

Newton was, without doubt, Galileo's great successor as a physicist – the idea that one was born in the year the other died is pleasingly suggestive of the passing on of a baton.

Yes. It's true or false (or both) that one was born the year the other died

I'm going to be strict here and won't accept either 'true' or 'false'. To get the point, you need to answer either 'yes' or 'both'.

It is commonly said, as if it has some kind of cosmic significance, that Newton was born in the same year that Galileo died. And this was true at the time *in England*, though not in Galileo's Italy. Galileo died at the start of 1642, while Newton was born on Christmas Day 1642. However, if we apply the Gregorian calendar (as they would have done back then in Italy), Newton was born the following year to that of Galileo's death.

What is certainly incorrect is to celebrate Newton's birthday on the current Christmas Day. The calendar reform that resulted in Newton's birthday shifting a year had been due for a long time. In 1276, for instance, Roger Bacon had pointed out the problem in the old Julian calendar, which had leap years every fourth year without fail, and as a result gradually drifted out of synch with nature. As the date of Easter was based on a natural phenomenon (the Spring equinox), this meant that Easter would drift away from its 'correct date'.

The Gregorian calendar was finally introduced (in a form that was very similar to Bacon's recommendations) in parts of mainland Europe in 1582, but would not be accepted in the UK until 1752 (and didn't arrive in Russia until 1918). This means we simply can't say 'true' or 'false' for this statement, as the answer depends on the country from which you were making the observation.

Further reading: **The Calendar**

QUESTION 5
Saucer of secrets

Why are flying saucers called 'flying saucers'?

Answer overleaf ➜

While you're thinking ...

The more common term now for a flying saucer is UFO, for 'Unidentified Flying Object', although this is a strangely unsatisfactory term. Technically a bird, for instance, is a UFO until you identify it as such, and a genuine alien spacecraft would not be a UFO once its origins were established.

The world's favourite flying saucer, the Frisbee, dates back to the 1930s, when Fred and Lucile Morrison (then not married) were asked if they would sell a cake pan that they were throwing between them on the beach.

The Frisbee was originally called the 'Flyin-Saucer', with the later version, which is closer to the current one in design, called the 'Pluto Platter'. Students gave it the nickname 'Frisbee' after the Frisbie Pie Company, and it stuck.

They are called flying saucers because of the way they were said to move in an early report

I'm going to be generous and award you half a point if you say 'they aren't', because the term 'flying saucer' has been pretty much eclipsed by the more general-purpose 'UFO'.

There is some dispute over the origins of the term. The widely accepted version is that a US pilot saw some UFOs in 1947. These craft appeared to be crescent-shaped, but the pilot, Kenneth Arnold, said that they moved erratically in the sky, 'like a saucer if you skip it across a pond'. His words were picked on by the headline writers who saw an easy-to-grasp term and referred to 'flying saucers'.

Interestingly, after the wide media coverage, people started to report seeing saucer-shaped craft, which became the most common shape for a while, before different media reports favoured other shapes. Having said that, even though the sketch Arnold drew definitely wasn't saucer-shaped, he was quoted as saying that the UFOs looked like saucers or discs – but given the reporting standards of the time, this could easily have been transplanted from his remark about their motion.

The erratic movement, seeming to defy the laws of physics, is often portrayed as proof that the UFOs have some unearthly technology, as a plane could not behave in this way, though more sceptical observers suggest that this demonstrates instead that what is being seen is an optical effect, which can easily dance around in an uncanny fashion.

Further reading: **The Universe Inside You**

Blue heaven

Why is the sky blue?

Answer overleaf ➔

While you're thinking ...

The surface temperature of the Sun is around 5,500°C,
which means it is an emitter of white light.

The air around us on Earth consists of 78 per cent nitrogen,
21 per cent oxygen, just under 1 per cent argon – by which
time there is less than 0.04 per cent for the other gases.

Most of that 0.04 per cent is carbon dioxide.

The sky is blue because the air scatters blue light more than reds and yellows

When the white light from the Sun, which is a mix of all colours of the spectrum, hits gas molecules in the air, some of the photons are absorbed and re-emitted in a different direction, a process called scattering. But some colours are absorbed far more than others, notably the blues. The result is that the scattered light gives the air an apparent blue colouration.

With a significant portion of the high-end blue light removed, the light coming to us directly from the Sun appears yellowish, which is why the Sun is almost always depicted as yellow, rather than the more accurate white. As evening draws on and the Sun sinks lower in the sky, the light passes through more of the atmosphere before it reaches your eyes. This means that more of the light is scattered and the result-ant light directly from the Sun is weaker and more towards the red end of the spectrum.

The blue colour from scattering was first observed experi-mentally in the late 19th century, when light was shone through clear liquid with tiny particles suspended in it, which led to the supposition that the blue sky was also caused by particles, whether dust or tiny water drops. However, if this were true, you would expect significantly more blueness when the air is heavily polluted, which doesn't happen. It was only around 1911 that the correct explanation was shown to be mathematically correct and expected to produce the blue skies we all enjoy.

*Further reading: **Light Years***

Ray gun remembered

Who is said to have devised a ray gun around 212 BC?

Answer overleaf ➜

While you're thinking ...

The first known use of the term 'ray' in English to denote a beam of light was in the 14th century, in the poem *Pearl*, where we read: 'A crystal clyffe ful relusaunt / Mony ryal ray con fro hit rere' which roughly translates as 'A crystal cliff that shone full bright / Many a noble ray stood forth.'

One of the earliest examples of fictional ray weapons was in Washington Irving's 1809 book *Conquest of the Earth by the Moon*, where the technologically advanced aliens were 'armed with concentrated sunbeams'.

A more specific early example of a death ray was in H.G. Wells' still very striking novel from 1898, *The War of the Worlds*. The Martians' walking war machines emit 'heat rays', which are powerful beams of infrared light, concentrated by a parabolic mirror.

Archimedes is said to have invented a ray gun

Archimedes lived in Syracuse on Sicily, which was conquered by the Romans around 212 BC. He designed a number of innovative defence engines, including a collection of curved metal mirrors that would concentrate the Sun's rays with the intention of setting an attacking ship on fire.

As far as we know, these mirrors were never built (quite possibly because no one believed they would work). Even today it's not entirely clear whether the concept was practical. The TV show *MythBusters* has made several attempts to recreate the mirror weapon, each of which has failed. However, a number of academics have cast doubt on the TV 'experiments', and other trials have succeeded in starting some flames.

It seems likely in perfect weather conditions, with very stable mirrors, that relatively flammable material on deck (as opposed to wet ship's planking) could have been set alight.

In a sense, the dispute is pointless, as there is no evidence that the mirrors were ever made. But if they had been, at the very least the intense beams would have been terrifying – and quite possibly blinding – for the sailors. And it is a mark of Archimedes' inventiveness that he conceived of the ray gun long before it became a trademark of science fiction.

Further reading: **Ten Billion Tomorrows**

Find the link

What have Lucian of Samosata, Francis Godwin, Johannes Kepler, Cyrano de Bergerac and Jules Verne in common?

Answer overleaf ➜

While you're thinking ...

Lucian of Samosata was a Graeco-Roman satirist who is best known for *A True Story*, one of the first novels, which appears to be a parody of the *Odyssey*.

Francis Godwin was an Anglican bishop in the early 17th century, while Kepler was a German astronomer and mathematician from the same period who established the (observationally derived) laws of planetary motion.

Cyrano de Bergerac was a real person, a French dramatist in the first half of the 17th century. He is best known from the (mostly fictional) play of his life, written by Edmond Rostand, a younger contemporary of the great French novelist credited as one of the key figures in the development of science fiction, Jules Verne.

They all wrote fictional accounts of a trip to the Moon

Long before true science fiction, trips to the Moon were used for allegorical purposes or in strange fantasy tales. For Lucian this was primarily a vehicle to mock the *Odyssey*, but Godwin's tale was more interesting, as it came at a time when the nature of the Moon was still in dispute. Was it the perfect sphere that Aristotle's 'science' required it to be, or was it, as Galileo suggested, a body like the Earth with mountains and seas?

Godwin didn't attempt anything vaguely scientific to get his narrator there (perhaps as a defence against contradicting Aristotle, the book is supposedly written by a Spaniard called Domingo Gonsales). He is pulled by a flock of special swans called gansas that migrate to the Moon each year, who take him to a very Earth-like environment. Nor, for that matter, does Kepler use scientific means. His hero crosses an insubstantial bridge of darkness, used by demons.

Cyrano attempts a kind of scientific approach, first trying to use the lifting power of the Sun on dew, then, with apparent prescience, rockets. But true science had to wait until Verne came along. At least, sort of true science. Verne infamously criticised the younger H.G. Wells for using an imaginary substance, cavorite, to get his travellers to the Moon in his book *The First Men in the Moon*. But in reality, once you get past the imaginary substance, at least Wells' approach would have worked, while the cannon-powered ship that Verne employed in *De la Terre à la Lune* (*From the Earth to the Moon*) would have left his astronauts as a splattered puddle under the vast g forces required to get a ship to escape velocity in the length of a gun barrel.

*Further reading: **Ten Billion Tomorrows***

HOW MANY MOONS DOES THE EARTH HAVE?

QUIZ 2
ROUND 5: SPACE

Travelling light

If you travel towards a source of light at high speed (say half the speed of light) how will the light's speed and colour change?

Answer overleaf

While you're thinking ...

Light is a self-supporting interaction between electricity and magnetism.

Scientists who thought light was a wave invented an invisible material called the ether, that filled all space, to explain how it could pass through a vacuum.

Travelling at half the speed of light, you would get from London to New York in less than 4/100ths of a second.

The light's speed will be unchanged, but its colour will shift towards the blue

If, instead of the light, it were a truck heading towards you at, say, 100mph, and you headed towards it at half its speed, the truck would now be heading towards you at 150mph. This is Galileo's version of relativity, where velocities simply add together. But Einstein realised that light behaves totally differently from normal moving objects. His special theory of relativity makes the assumption that however fast you move away from, or towards, a light source, the light still travels onward at the same speed. This is because the interplay between electricity and magnetism that makes up light will work only at a particular speed. If light changed speed when you moved with respect to it, it could no longer exist and would disappear.

But something does happen when you move relative to the light source. Like the more familiar Doppler shift that happens with sound, causing a siren to drop in pitch as it passes by, if you move away from a light source, its frequency drops – it heads towards the red end of the spectrum (a red shift). If you move towards a light source, its frequency goes up. It undergoes a blue shift. If, for instance, you started with a mid-range red light, which has a wavelength of around 680 nm (frequency around 440 THz), the result of moving towards it at half the speed of light would be to shift it all the way to a wavelength of 390 nm (frequency around 760 THz), putting it at the high end of the violet range, just about to disappear into the ultra-violet.

Further reading: **Light Years**

QUESTION 2
Seeing stars

Does the fact that the night sky is mostly black and not full of stars in all directions prove that the universe is not infinite?

Answer overleaf

While you're thinking ...

The visible universe stretches about 45 billion light years in each direction – but the whole thing could be much bigger.

The brightest star seen from Earth is Sirius
in the constellation Canis Major.

The Atacama Desert in Chile is generally considered to have the most perfect 'dark' night sky on Earth, hence the location of a number of major telescopes there.

No, a black sky does not prove that the universe is finite

We don't know if the universe is finite or infinite (and if it's finite, just how big it is). And we can't deduce anything from the blackness of the night sky. If there were an infinite set of stars out there, then whatever direction you look in, you should eventually hit a star, and you might therefore expect the whole sky to glow with starlight. Known as Olber's paradox, after the 18th-century astronomer Heinrich Wilhelm Olber, there have been a number of arguments explaining why this isn't the case.

Olber favoured the idea that it was interstellar dust that caused the darkness of the sky, blocking most of the more distant starlight from ever making it to our eyes. Others have suggested that the missing stars are the work of red shift. We know the universe is expanding, and the further away bodies are, the faster they are receding. When an object moves away from us, the wavelength of the light that it gives off shifts towards the red end of the spectrum. With a big enough red shift, visible light shifts into the infrared and becomes invisible to our eyes.

This is a real effect, but not the main reason we would have a mostly black sky, even in an infinite universe. The answer was first suggested by Edgar Allan Poe in his *Eureka: a prose poem*, an essay he wrote from a lecture on cosmology he gave in New York in 1848. Poe pointed out that in a universe with a finite age, light from the more distant stars would not have time to reach us. It's perfectly possible that there are stars, clusters and galaxies out there in the directions where the sky appears black – but we can't see them yet.

*Further reading: **Before the Big Bang***

Universal knowledge

What percentage of the universe (in mass/energy terms) is made of stuff we understand?

Answer overleaf ➡

While you're thinking ...

Mass and energy are conveniently interchangeable, as described by Einstein's beautiful equation (arguably the world's most famous), $E = mc^2$.

The universe is thought to be around 13.7 billion years old.

The visible universe – the bit we can see – is around 45 billion light years in each direction. This is because the universe has expanded since the light set off towards us.

Around 4.9 per cent of the universe is made from stuff we understand

If you said, 'None of it, because it's based on quantum particles, and as Feynman said, no one understands them', you get a pat on the back, but no points. You can have a point for 4 or 5 per cent, as both are frequently used.

This is a reference to two surprising components of the universe. Around 26 to 27 per cent is thought to be dark matter. This is a hypothetical different kind of matter that is influenced by gravity, but not electromagnetism. It was dreamed up to explain the way large bodies like galaxies and groups of galaxies behave.

When something spins around, its component parts try to carry on in a straight line. It's only a centripetal force (gravity, in the case of galaxies) that holds it together. But spin it fast enough and it should break apart. Big celestial bodies spin so quickly that they should do just that – so the idea is that there is extra, invisible mass, holding them together. There is an alternative theory, Modified Newtonian Dynamics, which suggests that bodies of this size don't behave quite the same as normal-sized objects. If this were the case, we wouldn't need dark matter.

The remaining 68 per cent is labelled dark energy. This is the repulsive force causing the expansion of the universe to accelerate. Although tiny locally, over the entire universe it requires a vast amount of energy, and as mass and energy are equivalent, it can be thought of as a large percentage of the universe. So, if this is all correct, the bit we know, the universe we can see and measure, is just 4 to 5 per cent of the entirety of the universe.

Further reading: **The 4% Universe**

Ill-met by moonlight

How much brighter is sunlight than full moonlight?

Answer overleaf ➜

While you're thinking ...

The Sun is around 1.4 million kilometres (870,000 miles) across, compared to the Moon's 3,470 kilometres (2,156 miles).

The light of the full Moon is bright enough to read by and cast shadows, though not bright enough to trigger full colour vision.

Although the 'dark side of the Moon' is often referred to (hands up, Pink Floyd), it doesn't exist. The far side, which always faces away from us, gets plenty of sunlight.

Sunlight is typically between 300,000 and 500,000 times brighter

Give yourself a full mark for anything between 300,000 and 500,000 and half a mark for anything outside that between 200,000 and 600,000.

We don't really get a feel for the vast range of illumination between sunlight and moonlight because of a combined effort from our eyes and brains. In moonlight, the iris of the eye has relaxed to allow the pupil – essentially a hole to let light through – to get bigger. This means that more light comes in. But a far bigger effect is the way your brain processes the image that you see.

Because the 'view' we see is a construct of the brain, objects can appear far brighter or dimmer than they really are. Rather like those music systems that keep the sound level appropriate to deal with the range of sound in a particular track, so the brain sets its concept of light and dark according to the overall level of light coming in. There is still a very clear visual difference between sunlight and moonlight, but the apparent variation is nothing like that real factor of up to half a million.

We see the full Moon as shining with something between a yellow and a blue-white sheen, depending on where it is in the sky. (When it's lower, the light has to pass through more air, where blue light gets scattered more than red, so it appears yellower when near the horizon.) But the actual surface of the Moon, illuminated by sunlight, is a dirty grey. It's only by contrast with the darkness of the night sky that it appears so bright.

*Further reading: **Light Years***

QUESTION 5
Weighing up the solar system

To the nearest 1 per cent, what percentage of the mass of the solar system is in the Sun?

Answer overleaf ➜

While you're thinking ...

Jupiter has a mass of approximately 1.9 × 10²⁷ kilograms, about 318 times the mass of the Earth.

The Moon is the fourteenth most massive body in the solar system. Among all the moons of the solar system, only Ganymede, Titan, Callisto and Io are heavier.

The Earth is easily the most massive of the 'rocky' inner planets, around 23 per cent heavier than Venus.

99 per cent of the solar system's mass is in the Sun

Although Jupiter is much heavier than the Earth, the Sun is in a different class, at 1,047 times more massive than Jupiter and around 300,000 times more massive than the Earth.

As stars go, the Sun is fairly unimpressive. For example, the white supergiant star Rigel, which is easily spotted as the bottom right star of the main seven stars in the constellation of Orion, is estimated to be at least twenty times the mass of the Sun, while the star R136a1, which is in the Large Magellanic Cloud, is the most massive known star, around 265 times the mass of the Sun.

At the bottom end of the stellar scale are brown dwarfs, which aren't really stars at all, because they don't have enough material in them to generate the temperatures and pressures required for hydrogen fusion, so can't burn like a conventional star. Brown dwarfs have masses between around fifteen times that of Jupiter and 80 times, though the exact borderlines between a giant planet, like a super-Jupiter, and a brown dwarf are not distinct.

The Sun is technically a 'main sequence' or dwarf star, sitting in the middle of the Hertzsprung–Russell diagram that shows the different families of stars. Rather confusingly, white dwarf stars, for which the term 'dwarf' is much more appropriate, are not true dwarf stars, but the remains of stars between about eight and ten times the mass of the Sun, which have shed their outer layers and left behind something that is about the size of the Earth with a mass similar to that of the Sun.

*Further reading: **Stars: A Very Short Introduction***

QUESTION 6
Fast folk

How long would it take, travelling at the speed of the fastest vehicle to carry humans (so far), to cover one light year?

Answer overleaf ➜

While you're thinking ...

Voyager 1 is the fastest-travelling human-made device at around 17 kilometres per second with respect to the Earth, but a manned vehicle hasn't travelled this quickly.

Concorde, the fastest vehicle that ordinary passengers have travelled on, flew at Mach 2.03, which is around 2,150 kilometres per hour (1,340 miles per hour) at its cruising height.

The nearest star, other than the Sun, is over four light years distant.

At Apollo 10's speed it would take 27,000 years to cover a light year

There are two parts to the question – allow yourself half a point if you knew that the fastest humans have travelled was in Apollo 10, and another full point if you are within 500 years either way of the journey time.

Apollo 10's speed is, indeed, the fastest humans have travelled (relative to the Earth), reaching an impressive 39,897 kilometres per hour (24,790mph). But light makes this look pretty pathetic at 299,792,458 kilometres per *second*.

This obviously makes a light year – the distance light travels in one year – impressively large. And at Apollo 10's speed, it would take 27,000 years to cover one light year. Think how much human civilisation has changed in that time period.

When you realise that the nearest star is around four light years away, and the nearest star so far discovered that might have a habitable planet is at least twelve light years away, the sheer scale of interstellar distances becomes apparent. If we are ever to travel to the stars, we need to move a whole lot faster than we have so far.

Further reading: **Final Frontier**

Venusian values

What's the average surface temperature on Venus?

Answer overleaf ➔

While you're thinking ...

Venus is the nearest planet in size to the Earth, and though being nearer the Sun, was once thought to have a tropical climate.

The surface of Venus had never been observed before probes arrived, as it has permanent 100 per cent cloud cover.

The average surface temperature on the Earth is 14°C (57°F).

The average surface temperature on Venus is 460°C (860°F)

Have a point for anything between 440°C (824°F) and 480°C (896°F).

It was something of a shock when the first successful probe, Mariner 2, reached Venus in 1962 and started to send back data that uncovered the reality of this hellish planet. Not only do surface temperatures on Venus average around 460°C, they can reach as high as 600°C. It's significantly hotter than Sun-hugging Mercury. The metal lead would run liquid on the surface of Venus.

The planet does have an atmosphere, one of the reasons it was thought to be a hopeful home for life – but it is nothing like our atmosphere. Venus is swathed in a thick layer of carbon dioxide, giving it an immense atmospheric pressure at the planet's surface, over 90 times the pressure on Earth. The over-the-top greenhouse effect that all the Venusian carbon dioxide produces is responsible for those incredibly high temperatures. It's not the CO_2 that we see, though – carbon dioxide is transparent. As if to emphasise just how truly unpleasant Venus is, those bright white clouds that totally conceal the surface and make it so bright are sulphuric acid.

It's a strange day there too, because of the way Venus rotates, which is in the opposite direction to the Earth and most of the other planets. It takes Venus 243 (Earth) days to make a complete rotation, which is longer than the 225 days it takes to get around the Sun. But because of the way the planet spins, combined with the movement around the Sun, the time between sunrises is just under 117 days.

*Further reading: **Planets***

206   HOW MANY MOONS DOES THE EARTH HAVE?

QUESTION 8
A singular singularity

Name the location in the universe where the Big Bang started

Answer overleaf ➜

While you're thinking ...

The Big Bang was given its name by astrophysicist Fred Hoyle
in a BBC radio broadcast and was a term of mild derision.

Arguably there has never been a less apt name, as
the Big Bang wasn't big and didn't bang.

The pre-Big Bang 'seed' of a universe, in the earliest version of
the theory by Georges Lemaître, was referred to as a 'cosmic
egg', sometimes rather clumsily contracted to 'cosmeg'.

Any location is correct for the start of the Big Bang

Hold your finger up and point at a space in front of your nose – this is where the Big Bang happened. I can say that with confidence without knowing where you are, because you can do this in any part of the universe and it holds true.

According to the Big Bang theory, which remains the best-accepted theory for the origins of the universe at the time of going to press, the entire universe began as an infinitely small point or singularity. An extremely short time afterwards, the nascent universe began expanding in the Big Bang. (So, strictly speaking, the universe did not begin with the Big Bang in the Big Bang theory.)

If this had been a normal explosion, out into space, then we could look back and say that the Big Bang happened at a particular location, just as we can say that an explosive charge in a quarry went off in a particular location, even though, in the process, it brought down a 500-metre section of the quarry's face.

However, the Big Bang was anything but normal. This was not a case of mass/energy from the Big Bang expanding into existing space, but rather that space itself was expanding from pretty much nothing. And that means that every and any point inside the universe is located 'where the Big Bang happened' because every bit of the universe was at that point when it happened.

Interestingly, because space was expanding, it could do so far faster than the speed of light, which is only a limit *within* space – without such a rapid expansion (known as inflation), the Big Bang theory would not fit the observed universe.

Further reading: **Before the Big Bang**

QUIZ 2
ROUND 6: TECHNOLOGY

Charge your weapons

What does 'Taser' stand for?

Answer overleaf ➜

While you're thinking ...

The Taser is a weapon that is intended to disable with an electric shock, produced by US company Taser International Inc.

Most Tasers fire darts with attached wires that send a charge from the device to the victim's body, disrupting muscle control and temporarily disabling the individual. One product, used in the controversial Raul Moat case in the UK, packages the entire Taser system into a shotgun cartridge and so works wirelessly.

The Taser was originally developed by Jack Cover, an ex-NASA employee, in around ten years leading up to its launch in 1974.

'Taser' stands for 'Thomas A. Swift's Electric Rifle'

The obvious origin of the name is as a derivative of 'laser' or 'phaser' – both terms for weapons using an insubstantial beam. The phaser is fictional from *Star Trek*, but the laser (standing for 'light **a**mplification through **s**timulated **e**mission of **r**adiation') is of course real, if unable to provide the kind of non-lethal stunning associated with a phaser.

Lasers had been invented only six to seven years before Jack Cover began work on his weapon – and it still seems likely that the laser name was an inspiration. But Cover has always insisted that he took the name from a favourite childhood book, *Tom Swift and His Electric Rifle*.

In the book, Tom's amazing rifle delivers a remote electrical charge (as some sort of beam – there are no wires, and it can pass through a wall) to do anything from stun a human to kill an elephant. So it certainly does seem an appropriate inspiration. Cover did fudge one aspect, though. Tom Swift is never given a middle name in the book, so that 'A.' is purely there to get the desired '-aser' format.

Tasers remain controversial. They no doubt save lives when deployed where otherwise someone would be shot dead (the Taser International website has a rolling 'lives saved' counter), but Tasers tend to be used significantly more frequently than conventional firearms would be, particularly in countries that are less liable to use guns than the US, which is most countries where they are deployed. Despite being intended to be non-lethal, there is no doubt that Taser use has resulted in a number of deaths.

Further reading: **Ten Billion Tomorrows**

QUESTION 2
Wizard words

**What was significant about these words:
'YOU ARE STANDING AT THE END OF A ROAD
BEFORE A SMALL BRICK BUILDING. AROUND
YOU IS A FOREST. A SMALL STREAM FLOWS
OUT OF THE BUILDING AND DOWN A GULLY'?**

Answer overleaf ➜

While you're thinking ...

The capital letters in this question have a kind of significance.

Green on black would probably be the best colouring for this
text, though it also works as printed black on white paper.

These words were first seen in 1976.

They were the introductory lines of *Adventure*, the first computer adventure game

Inspired by playing *Dungeons and Dragons*, which came out in 1974, computer engineer Will Crowther, a caver in his spare time, put together the first text-based computer adventure game, set in Colossal Cave. The game would soon be improved by grad student Don Woods and made available on a wide range of the computers of the day.

It's hard to recapture the excitement that players (like me) felt, hunched over a keyboard in a darkened computer lab. If you typed GO IN, you would be told:

YOU ARE INSIDE A BUILDING. A WELL HOUSE FOR
  A LARGE SPRING.
THERE ARE SOME KEYS ON THE GROUND HERE.
THERE IS A SHINY BRASS LAMP NEARBY.
THERE IS FOOD HERE.
THERE IS A BOTTLE OF WATER HERE.

… and the adventure was under way.

Further reading: **Ten Billion Tomorrows**

QUESTION 3
Spot the bot

Why were robots androids and androids robots?

Answer overleaf ➔

While you're thinking ...

Robots are now widely used in industry, but are very different from
the science fiction vision of an intelligent, humanoid device.

Although there have been impressive humanoid robots like
Honda's Asimo, to date they have been relatively limited,
needing considerable bespoke programming to carry
out abilities like dancing or walking down stairs.

The word 'android' comes from Greek roots, meaning man-like.

Because the first things called 'robots' were what we now call 'androids', and vice versa

The word 'robot' is taken from Karel Čapek's play R.U.R. (standing for Rosumovi Univerzální Roboti, or Rossum's Universal Robots), which was first performed in 1920.

The modern distinction between robots and androids is that a robot is a mechanical device, most commonly humanoid in fiction, though real robots are more frequently single-purpose devices, like the robotic arms in a car factory or a robotic vacuum cleaner. By contrast, an android is a humanoid artificial life-form that is biological in nature. But in origin these definitions were reversed.

Čapek's robots (taken from the Czech 'robota', meaning forced labour) were biological constructs, human-like, created to perform work. By contrast, the first use of the term 'android' (or to be precise *androide*) turns up in *Chambers' Cyclopaedia* back in 1728, referring to purely mechanical constructions.

Even in relatively modern usage, there has been confusion. The character Data in *Star Trek: The Next Generation* is referred to as an android, and yet seems to be electronic and mechanical when opened up. Generally speaking, though, the convention does now apply that robots are mechanical and androids biological.

Further reading: **Ten Billion Tomorrows**

Casting light on lifeguards

What might an optics expert refer to as the '*Baywatch* principle'?

Answer overleaf ➔

While you're thinking …

Baywatch was an American TV drama, set on the beaches of Los Angeles and starring David Hasselhoff and Pamela Anderson among a stable of red-clad lifeguards.

The series first aired in 1989.

Baywatch was famous for its slow-motion scenes of lifeguards running down the beach, carrying a rescue float, as they headed for a swimmer in difficulty.

The '*Baywatch* principle' refers to the principle of least time

You can also have the point for the principle of least action. The actual '*Baywatch* principle' (even if it's not actually called this) is a bit of common knowledge among beach lifeguards. If you see someone in trouble in the sea, the quickest way to get to them is not usually to swim straight towards them. It's only straight towards them if they are immediately in front of the lifeguard. Otherwise, the quickest approach is to run further along the beach to cut down the distance travelled through the sea. This is because even the best lifeguards can run much faster than they can swim.

The reason this also applies in optics can be seen with what happens when, say, a beam of light travelling at an angle through the air hits a block of glass. As it passes into the glass, the beam bends inwards, towards a perpendicular line straight into the glass. This means that the beam has travelled further in air and less far in glass than it would have if it went in a straight line from its start to end point.

The speed of light is significantly lower in glass than it is in air. So the light has reduced the time it takes to get from A to B. In fact, when you work out the distances and combine them with the change in velocity, it turns out that light takes the path that minimises the time taken to get from A to B. This is known as 'the principle of least time'. But 'the *Baywatch* principle' is more fun.

Further reading: **Light Years**

Computer composition

Where was the first music generated by a computer (and for a bonus, in which year)?

Answer overleaf →

While you're thinking ...

One of the best-known examples of computer music (if faked) was the 'dying' computer HAL in *2001: A Space Odyssey*, singing 'Daisy, Daisy'.

Electronic music was transformed by the Moog synthesiser, designed by Robert Moog, which dates back to the 1960s, but these were analogue electronic devices, not digital computers.

The first commercial digital synthesiser dates to 1979.

The first computer-generated music was at Manchester University in 1950

A point for Manchester University and another for 1950 (exact year only).

For a long time it was accepted that the earliest computer-generated music was produced at Bell Labs in 1957, or sometimes it was thought to be using the CSIRAC device from around 1951, but it was in fact at Manchester in 1950.

Alan Turing had realised the potential of the computer to produce music somewhat earlier, and wrote a manual describing how to generate musical notes using the clicks that could be produced on the speaker, known as 'the hooter', that was built into the early Ferranti ACE computer to send error alerts. If the clicks were output rapidly enough, at thousands of times a second, the result would be a musical note corresponding to the frequency of the clicks.

Turing himself seems never to have gone further than producing the instructions on how to do this, but Christopher Strachey, working on the ACE at Manchester, took up the challenge and gave the first musical performance with 'God Save the King'. This later appeared on BBC radio, with the ACE also playing Glen Miller's 'In the Mood' and 'Baa, Baa, Black Sheep'. Oddly, the reporter the BBC sent was one of the presenters of *Children's Hour*, rather than a news or science journalist.

Further reading: **Turing**

QUESTION 6
Focusing on lenses

Why is a lens called a lens?

Answer overleaf ➜

While you're thinking ...

The oldest discovered lens dates back to around 700 BC.

Although Roger Bacon is often credited with inventing
spectacles, they are more likely to have originated
towards the end of the 13th century in Italy.

Both the telescope and the microscope came into use around the turn
of the 17th century, but it is difficult to be sure who first used them.

Because a lens is shaped like a lentil

The familiar word 'lens' is simply the Latin name for a lentil – a doubly convex lens is shaped rather like the pulse. The oldest recorded citation in English dates to 1693, but this reflects the fact that most scientific writing before this time was in Latin. The use goes back at least as far as Roger Bacon's great work of 1267, the *Opus Majus*, in which he suggests (in Latin) that one possible shape for the universe is 'lenticular' and goes on to explain that 'a lenticular shape is that of the vegetable called lentil'.

We don't know for certain, but it seems likely that Bacon experimented with crude predecessors of telescopes and microscopes, as he comments that lenses can be 'contrived so the most distant objects appear near at hand and vice versa ... For we can so shape transparent bodies, and arrange them in such a way with respect to our sight and objects of vision, that the rays will be refracted and bent in any direction that we desire, and under any angle we wish we shall see the object near or at a distance ... We may read the smallest letters at an incredible distance, we may see objects however small they may be, and we may cause the stars to appear wherever we wish.'

Still, it's interesting to think, next time you tuck into a plate of lentils in a vegetarian lasagne or enjoy a dhal from the local Indian takeaway, that you are eating a collection of tiny lenses in the form of *Lens culinaris*.

Further reading: **Roger Bacon**

QUESTION 7
Digital dreams

Where were the first computer animations produced?

Answer overleaf ➔

While you're thinking ...

The same computer that produced the first animation produced
the 3D wireframe model used in the movie *Alien*, which
won the 1979 Academy Award for best visual effects.

The most famous computer animation company, Pixar, now
part of Disney, started as the Graphics Group of Lucasfilm.

One of the earliest computer animations from the Pixar
team (still then with Lucasfilm) was the 'rebirth' of a
planet in the Genesis Effect used in *Star Trek II*.

The first computer animations were produced at the Rutherford Appleton Laboratory in Oxfordshire

In 1964, the Rutherford Appleton Laboratory in Oxfordshire opened the UK's first purpose-built computer laboratory to house one of the world's first supercomputers – the Ferranti Atlas 1. It cost around £3m (equivalent to around £80m 50 years later) and filled an entire two-storey building that was constructed to house it.

The Atlas processor used more than 5,600 circuit boards, which would have covered an area about the size of a tennis court – around 90,000 times bigger than a modern computer chip. One of its large magnetic disks could hold just two photographs.

The original Atlas Computer Laboratory established a national computing operation to support scientific research. The Atlas 1 was replaced after eight years of operation in 1973 with an ICT 1906A, but the centre continued to be called Atlas. The world's first computer animations, made at the centre, included an animated model of stress-loading across an M6 motorway bridge that was being built at the time.

The M6 bridge animation was the first entirely computer-produced engineering film to be made and won the Great Britain entry in the 1976 international Technical Films Competition in Moscow.

Further reading: http://www.chilton-computing.org.uk/acl/technology/atlas/p015.htm

K is for kilo, kitchen and klystron

How does a klystron link air traffic control and kitchens?

Answer overleaf ➜

While you're thinking ...

Air traffic control uses English as a standard for communication around the world – so a Japanese controller will speak to a Japanese plane in English. This is to ensure that instructions are understood by all aircraft in the area, whatever their country of origin.

The word 'klystron' is derived from the Greek *κλύζειν* (*klyzein*), meaning to wash or break over.

The first recorded use of the word 'klystron' was in 1939, in the *Journal of Applied Physics*.

A klystron generates microwaves, linking air traffic control and kitchens

A klystron sounds like something out of science fiction, but is in reality the name for a particular type of valve (vacuum tube) that was used both as a radio frequency amplifier and as an oscillator to produce microwaves. A familiar piece of lab equipment in the 1960s and 70s for microwave experiments was a '3 centimetre klystron'.

The klystron was used during the Second World War as the microwave generator for early radar. Originally called 'range and direction finding' by its British developers, 'radar' is a contraction of the American equivalent, '**ra**dio **d**etection **a**nd **r**anging'. Legend has it that radar was inspired by an attempt to investigate Nikola Tesla's claims to have produced an electromagnetic death ray. When radar was first developed and top secret, it is likely that the story of Royal Air Force pilots eating carrots to see better in the dark was propaganda to explain the ease with which they were finding their targets. Radar pumps out a beam of microwaves (low-frequency light, below the infrared), then detects any microwaves that are reflected off potential targets.

Microwave ovens, originally called 'radar ranges', were developed from radar technology. An American engineer, Percy Spencer, discovered by accident that a radar set he was working on melted a chocolate bar in his pocket. The technology used for microwaves is a 'cavity magnetron', a different type of cavity-based oscillator from the klystron. The cavity magnetron became more popular in Allied radar during the Second World War as it improved on the relatively low power of the klystrons used at the time and could work at shorter wavelength. Although you don't find klystrons in microwave ovens, the microwaves they produce link air traffic control and kitchens.

*Further reading: **Inflight Science***

QUIZ 2
FIRST SPECIAL ROUND: PERIODIC TABLE

Fill in the missing item (name, symbol or atomic number) for these ten elements:

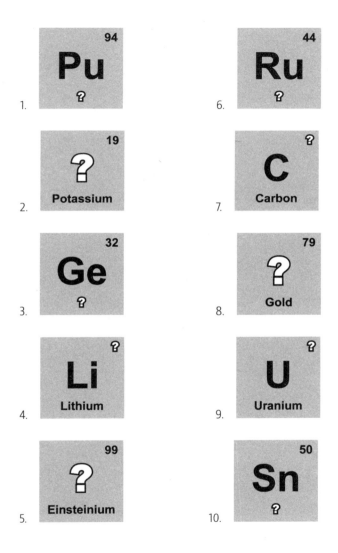

1. Pu 94 ?

2. 19 ? Potassium

3. Ge 32 ?

4. Li ? Lithium

5. 99 ? Einsteinium

6. Ru 44 ?

7. ? C Carbon

8. 79 ? Gold

9. ? U Uranium

10. 50 Sn ?

Periodic Table: Answers:

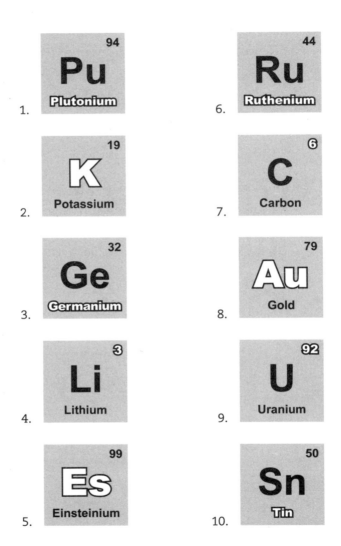

1. **94 Pu** Plutonium

2. **19 K** Potassium

3. **32 Ge** Germanium

4. **3 Li** Lithium

5. **99 Es** Einsteinium

6. **44 Ru** Ruthenium

7. **6 C** Carbon

8. **79 Au** Gold

9. **92 U** Uranium

10. **50 Sn** Tin

QUIZ 2
SECOND SPECIAL ROUND: NASA

Test your NASA knowledge:

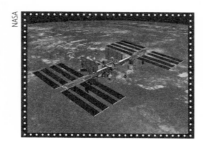

NASA

1. What is this?

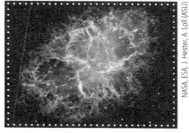

NASA, ESA, J. Hester, A. Loll (ASU)

2. What is the name of this remnant of a supernova, first seen in the year 1054?

NASA

3. What was the name of this type of capsule?

NASA

4. What is this satellite?

X-ray: NASA/CXC/SAO; Optical: Detlef Hartmann; Infrared: NASA/JPL-Caltech

5. What is the popular name of this galaxy, Messier 101?

6. Name either the third or fourth person to walk on the Moon.

7. What was the name of the 1960s NASA concept spaceship that was intended to surf on the shockwave of exploding nuclear bombs?

8. Which American physicist uncovered the O-ring problem that caused the *Challenger* shuttle disaster?

9. Which early US astronaut joked: 'They wanted to send a dog, but they decided that would be too cruel'?

10. What was the name of the first successful US satellite?

NASA: Answers

1. The International Space Station (ISS)

2. The Crab nebula

3. Gemini

4. The Hubble Space Telescope

5. The Pinwheel galaxy

6. Charles Conrad and Alan Bean (only one needed)

7. Orion (no relation to NASA's 21st-century Orion design)

8. Richard Feynman

9. Alan Shepard (after his Freedom 7 sub-orbital flight)

10. Explorer 1

FURTHER READING

If one of our topics catches your interest, here's a chance to find out more. Note that these books are not necessarily the source of the information in the quiz, but will allow you to read further around the topic.

30-Second Evolution, (eds) Mark Fellowes and Nicholas Battey (Icon Books, 2015)

A Brief History of Infinity: The Quest to Think the Unthinkable, Brian Clegg (Constable & Robinson, 2003)

Aspirin, Diarmuid Jeffreys (Bloomsbury Publishing, 2005)

Before the Big Bang: The Prehistory of Our Universe, Brian Clegg (St Martin's Press, 2009)

Build Your Own Time Machine [How to Build a Time Machine], Brian Clegg (St Martin's Press/Duckworth, 2011)

Diceworld: Science and Life in a Random Universe, Brian Clegg (Icon Books, 2013)

Ecologic: The Truth and Lies of Green Economics, Brian Clegg (Transworld, 2010)

Einstein: His Life and Universe, Walter Isaacson (Pocket Books, 2008)

Elephants on Acid, Alex Boese (Pan Books, 2009)

Gravity: How the Weakest Force in the Universe Shaped Our Lives, Brian Clegg (St Martin's Press/Duckworth, 2012)

How to Predict the Unpredictable: The Art of Outsmarting Almost Everyone, William Poundstone (Oneworld Publications, 2014)

If Dogs Could Talk: Exploring the Canine Mind, Vilmos Csányi (The History Press, 2006)

Inflight Science: A Guide to the World from Your Airplane Window, Brian Clegg (Icon Books, 2011)

Light Years: The Extraordinary Story of Mankind's Fascination with Light, Brian Clegg (Icon Books, 2015)

My Best Mathematical and Logic Problems, Martin Gardner (Dover, 2003)

Near-Earth Objects: Finding Them Before They Find Us, Donald K. Yeomans (Princeton University Press, 2012)

Nuclear Power: A Very Short Introduction, Maxwell Irvine (OUP, 2011)

Planets: A Very Short Introduction, David Rothery (OUP, 2010)

Roger Bacon: The First Scientist, Brian Clegg (Constable & Robinson, e-book edition, 2013)

Science for Life: A Manual for Better Living, Brian Clegg (Icon Books, 2015)

Stars: A Very Short Introduction, Andrew King (OUP, 2012)

Ten Billion Tomorrows: How Science Fiction Technology Became Reality and Shapes the Future, Brian Clegg (St Martin's Press, 2015)

The 4% Universe, Richard Panek (Oneworld, 2011)

The Calendar, David Ewing Duncan (Fourth Estate, 2011)

The History of Music Production, Richard James Burgess (OUP USA, 2014)

The Quantum Age: How the Physics of the Very Small Has Transformed Our Lives, Brian Clegg (Icon Books, 2014)

The Story of Measurement, Andrew Robinson (Thames & Hudson, 2007)

The Universe Inside You: The Extreme Science of the Human Body, Brian Clegg (Icon Books, 2012)

Think Like a Freak: How to Think Smarter About Almost Everything, Steven D. Levitt and Stephen J. Dubner (Allen Lane, 2014)

Turing: Pioneer of the Computer Age, B. Jack Copeland (OUP, 2012)

What If Einstein Was Wrong? Asking the Big Questions about Physics, (ed.) Brian Clegg (Ivy Press, 2013)

When Computing Got Personal: A History of the Desktop Computer, Matthew Nicholson (Matt Publishing, 2014)

The Quantum Age
How the Physics of the Very Small has Transformed Our Lives

'His best book yet'   John Gribbin
'Truly fascinating'   *Times Higher Education*

The stone age, the iron age, the steam and electrical ages all saw human life transformed by new technology. Now we are living in the quantum age, where quantum physics lies at the heart of every electronic device, every smartphone and laser. Quantum superconductors have made levitating trains and MRI scanners possible, and superfast, ultra-secure quantum computers will soon be a reality. And yet quantum particles such as atoms, electrons and photons remain enigmatic, acting totally unlike the objects we experience directly. Brian Clegg brings his trademark clarity and enthusiasm to a book that will give the world around you a new sense of wonder.

ISBN: 978-184831-846-5
£8.99

SCIENCE FOR LIFE
A MANUAL
FOR BETTER LIVING

AUTHOR OF THE BESTSELLING
INFLIGHT SCIENCE
BRIAN CLEGG

Science for Life
A Manual for Better Living

How do you make sense of the competing claims that surround almost every aspect of daily life? How can you really know what will work for you and make your life better? Hard science is the answer, and this manual is here to help you. Discover which 'superfood' contains carcinogens and 21 E-numbers, what does and doesn't enhance brainpower, and how your 'green' actions actually affect the environment. From toxins in food and nuclear power to exploring how best to stimulate the mind and avoid being manipulated, *Science for Life* is a reference guide like no other. Brian Clegg cuts through the vested interests and contradictory statements that litter the media to give a clear picture of what science is telling us right now about changing our lives for the better.

ISBN: 978-17857-802-57
£9.99